A Field Guide to the Raptors of Europe, the Middle East, and North Africa

D1549876

A Field Guide to the Raptors of
Europe, the Middle East, and North Africa

William S. Clark

Illustrations by N. John Schmitt
Foreword by Richard Porter

OXFORD
UNIVERSITY PRESS

OXFORD
UNIVERSITY PRESS

Great Clarendon Street, Oxford OX2 6DP

Oxford University Press is a department of the University of Oxford.
In furthers the University's objective of excellence in research, scholarship,
and education by publishing worldwide in

Oxford New York

Athens Auckland Bangkok Bogotá Buenos Aires Calcutta
Cape Town Chennai Dar es Salaam Delhi Florence Hong Kong Istanbul
Karachi Kuala Lumpur Madrid Melbourne Mexico City Mumbai
Nairobi Paris São Paulo Singapore Taipei Tokyo Toronto Warsaw
with associated companies in Berlin Ibadan

Oxford is a registered trade mark of Oxford University Press
in the UK and in certain other countries

Published in the United States
by Oxford University Press Inc., New York

A catalogue record for this book is available from the British Library

Library of Congress Cataloging in Publication Data

Clark, William S., 1937–
A field guide to the raptors of Europe, the Middle East, and North
Africa / William S. Clark; illustrations by N. John Schmitt.
Includes bibliographical references and index.
1. Falconiformes–Europe–Indentification. 2. Falconiformes–
Middle East–Identification. 3. Falconiformes–North Africa–
Identification. I. Title.
QL696.F3C625 1999 598.8′094–dc21 98-49178

ISBN 0 19 854662 9 (Hbk)
ISBN 0 19 854661 0 (Pbk)

Typeset by Best-set Typesetter Ltd., Hong Kong
Printed and bound in China

Foreword

by Richard Porter, *Bird Life International*

Gazing skywards in late September on the Bosphorus, during March at Eilat or in August at Falsterbo, you will never fail to be thrilled by the spectacle of raptor migration. Buzzards, eagles, harriers and hawks, often together with storks, soar ever upwards on air thermals before gliding off in lines to head for the next warm upcurrent that will lift them again. In this way, journeys of many thousands of miles between Europe and Africa can be sustained with hardly a wing-beat: energy conservation *par excellence*.

It was this spectacle that was instrumental in promoting the tremendous advancement in the field identification of birds of prey that has taken place in the last 30 years—an advance that has probably been more dramatic than for any other group of birds. Yet the band of birders that have been responsible for pushing the frontier ever forward has been surprisingly small. Prominent among them have been Dick Forsman and the author of this book, Bill Clark.

Bill, not content with his pioneering *Hawks*, covering the American raptors in the Peterson Field Guides series, has now turned his talents to the Western Palearctic. *A Field Guide to the Raptors of Europe, the Middle East, and North Africa* is a tribute to that talent being the culmination of years of painstaking observations and research.

Raptor identification can be regarded as an art form that has a useful function in the monitoring of what is happening in our environment. At the top of the food chain and relatively easy to count, raptors are often the first birds to indicate change, be it as a result of human persecution, habitat destruction, or environmental pollution.

But to provide a function as an indicator it is essential that the organism is correctly identified and, ideally, classified as to age and sex. It is an acknowledged fact amongst the 'Identification élite' that raptors are one of the most difficult group of birds to identify, especially the large falcons and the

Aquila eagles, and it is not surprising that in the late 1960s and early 1970s papers on migration were appearing in scientific journals that showed, incorrectly, Greater Spotted Eagles to be one of the commonest raptors on migration in eastern Europe (we now know these were Lesser Spotted Eagles and Steppe Eagles) and that Common Buzzards were uncommon (they were simply being mis-identified as Honey Buzzards).

It was at this time, in 1968 and 1970, that a group of Danes (Steen Christensen, Bent Pors Nielsen, N. H. Christensen, and L. H. Sorensen) published two papers on eagle and buzzard identification in *Dansk Ornitologisk Forenings Tidsskrift*. This was the first time that detailed and accurate line drawings had been attempted for the range of plumages of the difficult groups and this spearheaded the progress that raced to the end of the Millenium.

Bill Clark's book encapusulates that progress, covering the subject for the first time with colour illustrations and photographs, both essential accompaniments to an authoritative but user-friendly text. Both author and illustrator, John Schmitt, have given us a book that I'm sure will stand the test of time and give pleasure to thousands of raptor fanatics, as well as those who, after reading it from cover to cover, will still feel vulnerable when attempting to put a name to that black dot in the sky.

Acknowledgements

I first want to thank Richard Porter, Ian Willis, Steen Christensen, and Bent Pors Neilsen for their monumental work, *Flight identification of European raptors*, and also for many discussions and correspondence. This guide gave me a firm basis for first understanding and then continuing my research on the field identification of the diurnal raptors of the Western Palearctic.

Many people helped me during my initial research on raptors in Israel in the mid-1980s, especially Yossi Leshem, but also Ofer Bahat, Edna Gorney, and Ohad Hatzofe, and later Reuven Yosef. Also I acknowledge the contribution of many raptor banders: C. Schultz, K. Duffy, M. McGrady, M. Britton, C. McIntyre, T. Shohat, and A. Hinde.

Many people helped by taking me afield or providing information or hospitality or some combination of these; they are P. and C. Barthel, J. Bustamente, R. Clarke, A. Corso, A. Colston, J. and H. Eriksen, M. Ferrer, B. Heredia, N. Kjellin, T. Krueger, G. Leonoradi, K. Meyrom, B. Meyburg, R. Naoroji, J. J. Negro, I. Newton, R. Parslow, V. Prakash, D. Sargeant, A. Village, T. and V. Wehtje, and F. Zeisemer.

I thank Per Alstrom, Andrea Corso, Roger Clarke, Dick Forsman, Nils Kjellen, Frank Nicoletti, Hadoram Shirihai, Lars Svensson, T. Morioka, and others for stimulating discussions and correspondence on raptor field identification.

The following are thanked for making comments on draft accounts: first and foremost Andrea Corso, but also Per Alstrom, Chris Brown, Roger Clarke, Rob Davies, Nils Kjellen, Knut Niebuhr, Klaus Malling Olsen, and John Schmitt.

It was a great pleasure to work with John Schmitt, a most talented bird artist. Not only is he one of the best illustrators of raptors in the world, he was a delight to work with, as he was always cheerful and enthusiastic and took all constructive criticism well.

And, finally, thanks go to my sister, Ellie, for loads of encouragement and support.

Bill Clark

Artist's note

The reader will be mistaken to assume that, just because the illustrations are credited to one artist, they are the sole effort of one person.

Well before Bill Clark honoured me with the opportunity of collaborating with him on this project, I had been trying to satisfy my long fascination with the birds – and particularly the raptors – of Eurasia by collecting books, videos, etc., whenever I found (or could afford) them. Consequently I felt I was well prepared to embark on this project, and do justice to our goals. However, we hadn't progressed far into the work before I realised how inadequate my resources were.

Fortunately our project didn't hinge on my meagre resources. Bill Clark not only provided a vast collection of reference materials accrued over many years of work with the raptors of the Western Palearctic region, but I was also generously supplied with personal notes, reports, and photographs from many of his colleagues. They also helped mould the book by providing valuable comments on the plates and text. I regret that our acknowledgements will fall short of expressing our gratitude and the debt we owe to them all.

The art is also the product of considerable time spent in museum collections and in the field where notes on all aspects of plumage, soft parts, colours, and shapes, were collected. The value of museum skin collections as regards the development of my personal knowledge and how it helped define my work can hardly be overstated. In-depth examination of large series of skins, both before and after time spent in the field, greatly enhanced our understanding, and influenced the ways in which we observed and interpreted what we saw in the field.

However important the photographic and skin resources would prove to this project, they were only supportive compared to the inspiration and knowledge gained from time in the field. Those precious few days provided the motivations, insights, questions, and excitement that carried over into my notebooks. Soon my books brimmed with drawings from life – everything from simple pencil silhouettes to quick watercolour studies surrounded by written notes.

By combining all the above elements, and following the author's outlines and advice, I composed the figures on the plates.

Raptors, wherever they are found, often pose identification challenges to both novice and experienced bird enthusiasts because of their relative scarcity, variety of plumages, similarity *and dissimilarity* of silhouette. Our goal was to produce a book that picked up where the average field guide leaves off, allowing the user to go beyond those visible traits that may lead to satisfying conclusions.

I hope that you enjoy this book, and that it aids in elevating your level of confidence and appreciation for this dynamic group of birds and the varied landscapes that are their home.

Norwalk, California   N. John Schmitt
September 1998

Contents

Plates

Photographs of raptors

Photo credits

Most of the photographs used in the appendix were taken by the author. Those taken by other photographers are listed below:

Bearded Vulture 2	Dr. Larry Schwab
Hen Harrier 1	Roger Clarke
Hen Harrier 2	James White
No. Goshawk 1,2	Roy J. C. Blewitt
Spanish Imp. Eagle 1	Dr. Bernd-U. Meyburg
Eleonora's Falcon 5	Dr. Ludlow Clark
Merlin 1,2	Dennis Green
Merlin 4	Kent Carnie

Anatomy and plumage of a raptor

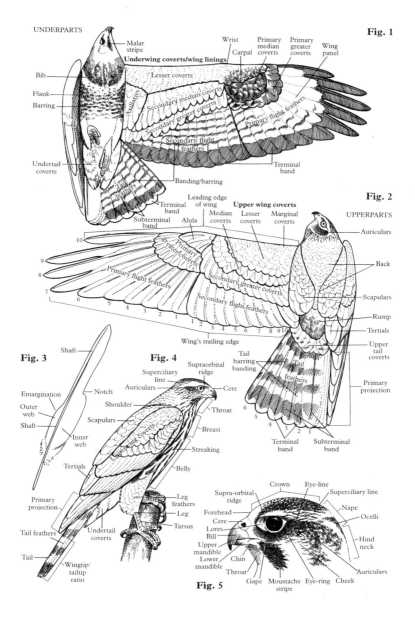

UNDERPARTS

Fig. 1

Bib
Flank
Barring

Malar stripe

Underwing coverts/wing linings

Lesser coverts

Wrist
Carpal

Primary median coverts
Primary greater coverts
Wing panel

Secondary median coverts
Secondary greater coverts

Primary flight feathers

Axillaries

Secondary flight feathers

Undertail coverts

Tail feathers

Banding/barring

Terminal band

Fig. 2

Leading edge of wing
Upper wing coverts

UPPERPARTS

Terminal band
Subterminal band
Alula

Median coverts
Lesser coverts
Marginal coverts

Auriculars

10
9
8
7
6
5
4
3
2
1

Primary greater coverts

Primary flight feathers

Secondary greater coverts

Secondary flight feathers

1 2 3 4 5 6 7 8 9 10

Wing's trailing edge

Back

Scapulars

Rump

Tertials

Upper tail coverts

Primary projection

Tail barring banding
Tail feathers

6
5
4
3 2 1

Terminal band
Subterminal band

Fig. 3

Shaft

Emargination
Outer web
Shaft

Notch

Inner web

Tertials

Primary projection

Tail feathers

Tail

Wingtip/tailtip ratio

Fig. 4

Superciliary line
Auriculars
Shoulder
Scapulars

Supraorbital ridge

Cere

Throat

Breast

Upper wing coverts

Streaking

Belly

Leg feathers
Leg
Tarsus

Undertail coverts

Fig. 5

Supra-orbital ridge
Forehead
Cere
Lores
Bill
Upper mandible
Lower mandible
Chin
Throat

Crown
Eye-line

Superciliary line
Nape
Ocelli

Hind neck

Auriculars

Gape
Moustache stripe
Eye-ring
Cheek

Glossary

Figure numbers refer to 'Anatomy and plumage of a raptor'.

adult plumage: the final breeding plumage of a raptor. Also called 'definitive basic plumage'.

alula: small feathered projection at the bend of a bird's upperwing.

albinism: rare abnormal condition when all of the feathers, beak, cere, talons, and eyes of a bird are without pigmentation. Feathers are white, and beak, cere, and talons are ivory. See **partial albinism** and **dilute plumage**.

auriculars: the feathers covering the ears (Figs 2, 4, and 5).

axillaries or **axillars**: feathers at the base of the underwing, also called the 'armpit' or 'wingpit' (Fig. 1).

back: see Fig. 2.

band: a stripe of contrasting colour, usually in tail. See **tail banding**.

barring: see Fig. 1.

beak: see **bill**.

belly: see Fig. 4.

bib: pattern of uniformly dark breast that contrasts with paler belly. Shown on Fig. 1.

bill: see Fig. 5; see also **mandibles**.

breast: see Fig. 4.

carpal: the underwing at the wrist, usually composed of all of the primary underwing coverts (Fig. 1).

cere: a small area of often colourful bare skin enclosing the nostril at the base of the upper mandible (Figs 4 and 5).

cheek: see Fig. 5.

chin: area directly under the beak (Fig. 5).

collar: a pale band across hind-neck.

coverts: the small feathers covering the bases of the flight feathers and tail both above and below (Figs 1, 2, and 4). See also **wing coverts** and **tail coverts**.

crown: top of head (Fig. 5).

dihedral: the shape when a bird holds its wings above the horizontal, further defined as:

    (1) strong dihedral: wings held more than 15 degrees above level;

    (2) medium dihedral: wings held between 5 and 15 degrees above level;

    (3) shallow dihedral: wings held between 0 and 5 degrees above level;

    (4) modified dihedral: wings held between 5 and 15 degrees above level; but held nearly level from wrist to tip.

dilute plumage: an abnormal plumage in which the dark colours are replaced by a lighter, usually creamy colour (but not white). See **albinism** and **partial albinism**.

emargination: an abrupt narrowing on the outer web of an outer primary. See also **notch** (Fig. 3).

eye-line: dark line through eye, often visible only behind eye (Fig. 5). Also called 'eye-stripe'.

eye-ring: the bare skin around the eye, somewhat wider in front of the eye in falcons (Fig. 5). Also called 'orbital ring' and 'scleral ring'.

face skin: the lores when bare of feathers.

facial disk: a saucer-shaped disk of feathers around the face, thought to direct sound to the ears. Most visible on harriers.

feather edge: the sides of a feather. Pale edges usually give the effect of streaking.

feather fringe: the complete circumference of a feather. Pale fringes usually give a scalloped appearance.

fifth winter: first adult plumage of raptors that take at least 4 years to reach adult plumage, especially large eagles, when some immature characters are noticeable.

flank: see Fig. 1.

fledgling: a raptor that has just left the nest, fledged.

flight feathers: the primaries and secondaries (Figs 1 and 2). Same as **remiges**.

forehead: see Fig. 5.

fourth winter: fourth non-adult plumage of raptors that take at least 4 years to reach adult plumage, especially large eagles.

gape: the mouth opening between the mandibles that parts when the bill is open.

glide: flight attitude of a bird when it is coasting downward. The wingtips are pulled back, more so for steeper angles of descent. Its tail is usually closed.

hackles: erectable feathers on the nape.

hawk: a raptor of the genus *Accipiter*, but also used generically for all diurnal raptors.

hind-neck: see Fig. 5.

hover: remain in a fixed place facing into the wind by flapping. More properly called 'wind hover'.

immature: non-specific term that means 'not adult'. It is used in some field guides to mean 'juvenile', and in others for the stages between juvenile and adult.

inner web of feather: see Fig. 3.

juvenile plumage: first complete plumage acquired in the nest. Usually differs from plumage of adults.

kite: remain in a fixed place in moving air on motionless wings.

leading edge of wing: see Fig. 2.

leg: see Fig. 4.

leg feathers: see Fig. 4.

length: distance from top of head to tip of tail.

lores: the area of the sides of face between the eye and the beak (Fig. 5).

lower mandible: see Fig. 5.

malar stripe: a dark mark on the cheek that begins at the base of the lower mandible (Fig. 1). Many buzzards show them.

mandibles: the upper or lower half of the bill (Fig. 5).

melanism: rare abnormal condition when a raptor's feathers are darker due to excess dark pigmentation.

monotypic: having no subspecies.

morph: term used for recognizably different forms of a species, usually colour-related. Colour morphs are dark, rufous, and light. See **phase**.

moult: means by which a bird replaces its feathers.

moustache stripe: a dark mark that begins under the eye, also called 'moustache mark' (Fig. 5). Most falcons show them.

nape: the back of the head (Fig. 5).

notch: abrupt narrowing of the inner web of an outer primary (Fig. 3). See also **emargination**.

ocelli: dark or light spots on the nape and hind neck that resemble eyes (Fig. 5).

outer web of a feather: see Fig. 3.

partial albinism: abnormal condition in which a few to almost all feathers of a raptor lack pigmentation and appear white. Feathers can be part white and part normal. Eyes, cere, and beak colours are usually normal; some talons may be ivory. See **albinism** and **dilute plumage**.

phase: term formerly used for colour morph. Phase implies a temporary condition; colour morphs are permanent. See **morph**.

plumage: all of the feathers of a bird collectively.

primaries: the outer flight feathers (Figs 1 and 2).

primary projection: the distance on the folded wing of perched birds from the tip of the longest primary to the tips of the secondaries. See Fig. 4.

raptor: any bird of prey; any member of the Falconiformes or Strigiformes, although sometimes used to refer only to the diurnal birds of prey.

rectrices: another term for 'tail feathers' (Figs 1 and 2).

remiges: another term for 'flight feathers' (Figs 1 and 2).

rump: the lowest area of the back (Fig. 2).

scapulars: row of feathers between back and upperwing coverts of each wing (Fig. 4).

scavenger: raptor that eats carrion, garbage, offal, and other animal waste.

second winter: immature plumage that follows juvenile on raptors that take more than one annual moult to reach adult plumage.

secondaries: the inner flight feathers (Figs 1 and 2).

shaft of a feather: see Fig. 3.

shoulder: see Fig. 4.

soar: flight attitude of a bird with wings, and usually tail, fully spread. Used to gain altitude in rising air columns.

spotting: round dark markings, usually on underparts.

streaking: vertical dark markings, usually on underparts (Fig. 4).

subadult plumage: plumage of a raptor that precedes adult plumage and appears much like it but with some immature plumage characters retained. See also **fifth winter**.

subterminal band: see Fig. 1.

superciliary line: contrasting line above the eye (Figs 4 and 5). Also called 'supercilium' and 'supercilia'.

supraorbital ridge: bony projection over the eye, giving raptors their fierce appearance (Figs 4 and 5).

tail: see Figs 1, 2, 4.

tail banding: see Figs 1 and 2.

tail coverts: feathers that cover the bases of the tail, above and below. See also **uppertail coverts** and **undertail coverts** (Figs 1 and 2).

talon-grappling: behaviour involving two flying raptors that lock feet and tumble with their wings extended.

tarsus: leg above the toes to the next joint. May or may not be covered by feathers (Fig. 4).

terminal band: see Figs 1 and 2.

tertials: innermost short secondaries (Figs 2 and 4).

third winter: plumage that follows second winter on large raptors that take more than 2 years to reach adult plumage.

throat: see Figs 4 and 5.

trailing edge of wing: see Fig. 2.

underparts: breast and belly (Fig. 4).

undertail coverts: see Fig. 1.

underwing: the underside of the open wing.

underwing coverts: see Fig. 1. Also called 'wing linings'.

upper mandible: see Fig. 5.

upperparts: back and upperwing coverts (Fig. 2).

uppertail coverts: see Fig. 2.

upperwing coverts: primary and secondary coverts (Fig. 2).

wing coverts: feathers that cover the bases of the flight feathers. Four sets are: marginal, lesser, median, and greater (Figs 1 and 2). See **upperwing coverts** and **underwing coverts**.

wing linings: underwing coverts (Fig. 1).

wing loading: weight divided by the wing area; a measure of the buoyancy of flight. The lower the wing loading, the more buoyant the flight.

wing panel: a light area in the primaries, usually more visible from below when wing is backlighted (Fig. 1). Also called 'window'.

wingspread: distance between wingtips with wings fully extended.

wrist: bend of wing (Fig. 1).

wrist comma: comma-shaped mark, usually dark, at the bend of the underwing; seen on the underwings of many *Buteo*.

Introduction

The aim of this guide is to present the latest information on tried and proven field marks for the field identification of diurnal raptors in Europe, the Middle East, and Northern Africa, henceforth designated as the Western Palearctic. Over 20 years ago, Porter *et al.* first published their monumental work, *Flight identification of European raptors*, which has been revised twice, most recently in 1981. That field guide was a milestone because, for the first time, the wing and tail shapes of flying raptors were shown correctly. Prior to this, no field guide, to my knowledge, showed the correct wing and tail shapes for raptors. Many field guides published since then have also failed to show accurately the shapes of raptors' wings and tails.

Although *Flight identification of European raptors* is a classic and provides a strong foundation, it does not cover perched raptors, contains only black and white drawings and photographs, and has plate captions that are hard to read quickly. Also, since its publication our knowledge has greatly increased by the many very good to excellent articles on raptor identification, plumages, behaviour, status, and distribution.

I decided to write this guide after more than 4 years of field work on raptors in Israel, where I came to know well almost all of the diurnal raptors of the Western Palearctic. Those that do not occur in Israel are familiar to me from North America. I also enhanced my familiarity with many species during extensive field work in India. I have coauthored a similar raptor field guide for North America in the Peterson field guide series, and a companion raptor photo guide.

The research to produce this guide was in five main areas. First, I took many photographs of raptors in the field, both perched and flying, as well as in zoos, in rehabilitation facilities, and hand-held raptors captured for ringing in many locations in Europe, Asia, and Africa. I also viewed many videos of raptors. Next, I studied specimens of raptors in numerous museums, both within and outside of the Western Palearctic. Third, I studied the literature on raptor identification and related subjects. Fourth, I studied the plumages of many raptors that had been captured for ringing or were in rehabilitation and took measurements on them to get more accurate data on wingspan, length, and weight. Finally, I spent much time afield in many locations field testing the field marks presented herein. Many of these field marks have been published by myself and others, but there is much new unpublished information reported here for the first time.

The system used here is based on the field mark system developed and used successfully by Roger Tory Peterson. The field marks presented here will enable a person who views a raptor in good light with good to excellent optical equipment to identify all but the most unusual individuals.

Colour plates

All of the 48 colour plates were drawn by John Schmitt specifically for this guide. The first four plates are summary plates, with many species illustrated for comparison. The remaining 44 are the species plates, with from one to three species illustrated per plate. Many species are shown on more than one plate (for example, Marsh Harrier males are on one plate, adult females and juveniles on another, and dark-morph birds on yet another plate). The plates are ordered by species in more or less taxonomic order; however, some species are placed out of order for space or comparison reasons. For example, Short-toed Snake Eagle was placed with Osprey, and Black-shouldered Kite with adult male harriers. The speciality plates for Arabia and Morocco are placed at the end.

The plates were prepared as a joint effort between the artist, John Schmitt, and myself. I provided John with draft species accounts (which included detailed descriptions of each plumage) for the species to be illustrated and short instructions on which plumages to show and which figures are to be flying from above, from below, or perched, front or back. I also provided him additional reference material such as colour slides or prints or both of raptors flying, perched, in hand, and in cages; colour and black and white photocopies of raptors; and articles on raptors to supplement his reference material. John then prepared an outline of the plate showing each figure in its exact size, shape, and position. I reviewed each outline for accuracy of wing and tail shapes and proportions among species and sex classes; often a few minor adjustments were made on the initial outline. After approval of the outline, John painted in the colour figures. John then sent me a colour copy of the plate. After any minor adjustments were made, John then painted in the pleasing backgrounds.

Colour photographs
Forty pages of colour photographs are included (pp. 287–367) to supplement the colour plates and text. Each photograph has a short caption. Only photographs of regularly occurring species are included.

Taxonomy
The taxonomy and order used were those of del Hoyo *et al.* (1994) except that the Barbary Falcon, *Falco pelegrinoides*, was treated herein as a full species and not as a race of the Peregrine, *F. peregrinus*.

Common names
The common names used are a combination of those used in Cramp and Simmons (1980) and del Hoyo *et al.* (1994), with some exceptions. Alternative common names are shown below the title in the species accounts.

Vagrants
All of the species of diurnal raptors that have occurred as vagrants in the Western Palearctic are included, except for Swainson's Hawk and Red-shouldered Hawk. The first is omitted because it has occurred only once, and the latter because of questionable identification, as the field marks used to separate the juvenile Red-shouldered Hawk from the juvenile Broad-winged Hawk were not provided in the sight record.

Species accounts

The main text of this guide consists of the species accounts, one for each species of diurnal raptor that has been recorded in the Western Palearctic.

Group heading

Preceding the accounts for each group of similar species is a group account, which may be a Family, in the case of the osprey; a genus, as for example, *Accipiter* and *Falco*; or a group of similar species not in the same genus, such as kites, vultures, or snake eagles. The species accounts for that group follow. The accounts are written in the following format.

Species accounts

These are composed of information under various headings and begin with the title of English common and scientific names, with pointers listing the colour plate number(s) and page number in the photo section for this species.

Description. This section begins with a descriptive sentence giving the type of raptor (e.g. buzzard), its relative size, and mentioning if it is a vagrant to the Western Palearctic. All field marks or general characters that apply to all individuals of this species, as well as differences in sexes and ages follow. Diagnostic field marks are given in italic type. Next are detailed descriptions of each regularly occurring plumage variation by age, sex, and colour morph, including eye, cere, and leg or toe colours. These descriptions are for raptors viewed close at hand.

Unusual plumages. Any of four abnormal plumages or cases of hybridization are included in this section. Abnormal plumage types are: albinism, partial albinism, dilute plumage, and melanism. Albinism is the complete absence of any pigmentation in the plumage, soft parts, and eyes and is very rare in raptors. In partial albinism, several to most or even all feathers lack pigmentation and are white, but the eyes and some soft parts are normal in colour. Dilute plumage is a condition in which the amount of melanin in many or all feathers is reduced so that normally brown feathers appear *café au lait* in colour. (This condition is also called 'leucism' and 'schizochroism'.) In melanism, there is an excess of dark pigment such that the raptor is completely dark brown on the body and coverts (but, interestingly enough, not always on the flight and tail feathers). This term applies only to abnormal plumages; when it occurs regularly, this coloration is called a 'dark colour morph'.

Similar species. All species that can be confused in the field with the species in each account are listed with the plate number or numbers on which each is illustrated, followed by the field marks to distinguish them. These are only given once; in the account of the other species 'See under that species for distinctions' is written following a short sentence on why they appear similar. In the accounts for vagrants, regularly occurring species that are similar are listed, but similar vagrant species are not listed in the accounts of the regularly occurring species.

Flight. Three methods of flight are described: how the wingbeats appear in powered flight and how the wings are held in soaring and gliding flight. If the species hovers, this is also noted. In some species, e.g. the kites, the particular flight methods that are distinctive are described.

Moult. This section is a description of when raptors moult their feathers. Some adults complete their moult annually on the breeding grounds before and after breeding (the moult is usually suspended while breeding). Others, especially trans-Saharan migrants, moult some feathers on the breeding grounds, but suspend moult during the autumn migration and complete their moult on the winter grounds. Some of the larger species do not replace all of their feathers in 1 year; usually 2 years or more are required to renew all feathers. The moult of juveniles is also discussed as it differs from that of adults in some species. Some begin moult in their first spring, when they are almost a year old, whereas others, particularly trans-Saharan migrants, begin moult earlier on the winter grounds and replace some or most of the body feathers and sometimes also a few tail and flight feathers. Moult is usually suspended during migration.

Age terminology is related to moults. All species are in juvenile plumage when they leave the nest. Many species undergo one annual moult into adult plumage; others, particularly the larger eagles, require three or four moults to reach adult plumage, with several immature plumages in between. See the 'Glossary' for the definitions of the immature plumages: 'second' through 'fourth winter', and 'subadult'.

Understanding the moult sequences of the flight feathers is important in order to to properly judge the ages of large raptors in immature plumages. (See Edelstam (1984), Miller (1941), and Jollie (1946) for a thorough discussion of moult in raptors.) In accipitrid raptors, primary moult begins with number 1, the innermost, and proceeds outward in sequence. In falconids, primary moult begins with number 4 and proceeds both inward and outward. When all primaries are not replaced in 1 year, then the sequence continues where it left off when moult is restarted the next spring. However, in accipitrids, moult of primary 1 occurs again during this moult season, beginning a new wave moult. Secondary moult of accipitrid juveniles begins at three moult centres, S1, S5, and S13 or 14. It proceeds inward from S1 and S5 and outward from S13 (also outward to replace the tertiaries). After the secondaries have been replaced once, the pattern of moult is apparently random.

Moult of the tail feathers is more irregular, however, usually beginning with T1, the central pair. The usual sequence is T2, T3, T6, T4, and then T5. However, there is lots of variation in this order and even some asymmetrical moult, particularly beginning with the second annual moult.

Behaviour. Any behaviours that will aid in field identification are described. Also described are hunting methods and the main prey. A brief discussion of nesting substrate and display flights follows.

Status and distributions. The breeding range or the permanent range and status within that range are presented, along with information on migration and dispersal. Any population declines are noted, with reasons for these given, if known.

Fine points. Any esoteric information on field marks or behaviour that will aid in field identification as to species, age, or sex are presented under this heading.

Subspecies. The races that occur in Western Palearctic of each polytypic species are given, along with the range of each race.

Etymology. The origins of the common and scientific names are given except for the obvious ones. In some case alternative explanations are given.

Measurements. Measurements of wingspread, total length, and weight are given. The range of values is given, with the mean in parentheses. Data on wingspread and total length in many references are incorrect; thus I took these measurements from live raptors whenever possible. This was supplemented by data taken from museum specimen tags, as they are for the most part almost the same as the data I took on live birds. The data on total length are taken from the top of the head to the tip of the tail, as this is how the raptors are seen when perched. These measurements are a bit smaller than those taken on specimens, which are from the tip of the beak with the head pointed upwards to the tip of the tail. Weight information in the literature is usually accurate; nevertheless, I used mostly data taken by me and my colleagues, supplemented by museum tag data, especially for those species that we did not handle as live birds.

Range maps

Range maps of the raptors covered are included in each species description to give a general idea of the range and distribution seasonally, with permanent resident, summer (not necessarily breeding), and winter ranges given. As migration paths between summer and winter ranges are usually broad, migrating raptors can be expected in the areas between them during spring and autumn migrations. Readers should consult regional and national bird handbooks for finer details of distribution.

Range maps are included in field guides to give the readers an idea of the geographic distribution of the birds covered. However, some readers read too much information into the maps; they fail to realize that these maps do not contain any habitat or density information for the species covered, nor do they tell the reader how obvious that species is in the proper habitat.

The range maps herein were prepared using as a base the maps prepared for the *Concise birds of the Western Palearctic* and updated with new range information from *The EBBC atlas of*

European breeding birds, Birds of Armenia, Birds of the Middle East, Les rapaces diurnes du Maroc, and from unpublished range maps of raptors in Georgia that were provided by Alexander Gavashelishvili.

How to use this guide

As a field guide, this book is to be carried in the field when birdwatching, especially if raptor watching. However, it is also to be used as a reference book, to be read and studied at home in preparation for field work. The reader should study the four summary colour plates to get a general feel for each type of raptor, concentrating on the overall shape of wings and tail but also noticing the field marks. One set of field marks serves to identify a perched raptor and another set applies to the same species flying. The summary plates do not cover perched raptors; these are covered on the other plates.

The guide contains a 'Glossary' where the terms used are defined and, in some cases, the reader is referred to one of the figures comprising the composite figure, 'Anatomy and plumage of a raptor', where the features of raptors are also indicated.

When a raptor is seen clearly and its field marks noted, even a beginner can use this guide to identify the bird correctly.

Some facts helpful in raptor field identification

- Juveniles in fresh plumage usually show pale tips to flight and tail feathers and greater wing coverts; the latter form narrow pale lines on wings. These tips usually wear off in winter.

- Non-juvenile raptors in summer and early autumn often show signs of moult in wings and tail or gaps in wings and uneven trailing edges of wings and tip of tail.

- The secondaries of many raptors are of different lengths in juvenile and older plumages. Some have longer secondaries as juveniles (the wings appear wider); others shorter ones (the wings appear narrower). Most raptors have longer tails in juvenile plumage.

- Flight and tail feathers are darker on uppersides as compared to undersides and darker on outer webs as compared to inner webs. Hence flight and tail feathers appear darker on the uppersides.

- Some raptors show pale areas on back-lighted underwings (windows or panels).

- Rufous underparts in fresh plumage of raptors, e.g. in juvenile Bonelli's Eagles, usually fade with exposure to sunshine and weather to buffy, creamy, or even whitish a few months after fledging.

- Raptors that show rounded wingtips when soaring often show somewhat pointed wingtips when gliding.

- There are two somewhat separate problems in raptor identification based on whether the bird is perched or flying. The field marks used in each case may be different. For example, wing shapes and the underwing patterns of soaring raptors are not visible on the same raptors perched, when the relative position of wingtip and tail tip can be field marks.

- When raptors are seen up close, much detail and shadings of colour are noticeable; the same raptor seen at a distance appears much more black and white, with loss of definition and colour.

- Raptors, and other birds as well, often appear different in colour under differing light conditions. All flying raptors appear darker against whitish skies, for example.

Any feedback would be appreciated and considered in future editions; write to Bill Clark c/o Oxford University Press.

Literature cited

Adamian, M. S. and Klem, D., Jr (1997). *A field guide to Birds of Armenia*. American University of Armenia. Oakland, CA and Yerevan, Armenia.

Bergier, P. (1987). *Les rapaces diurnes du Maroc*. Annales du CEEP, No. 3. Aix-en-Provence, France.

Cramp, S. and Simmons, K. E. L. (1980). *The birds of the Western Palearctic*, Vol. 2. Oxford University Press, Oxford.

del Hoyo, J., Elliot, A., and Sargatal, J. (ed.) (1994). *Handbook of the birds of the world*, Vol. 2. Lynx Edicions, Barcelona.

Edelstam, C. (1984). Patterns of moult in large birds of prey. *Ann. Zool. Fennici* **21**, 271–6.

Hagemeijer, W. J. and Blaire, M. J. (1997). *The EBBC atlas of European breeding birds*. T. & A. D. Poyser, London.

Jollie, M. (1947). Plumage changes in the Golden Eagle. *Auk* **64**, 549–76.

Miller, A. H. (1941). The significance of molt centers among the secondary remiges in the Falconiformes. *Condor* **43**, 113–15.

Porter, R. F., Willis, I., Christensen, S., and Neilsen, B. P. (1981). *Flight identification of European raptors*, (3rd edn). T. & A. D. Poyser, London.

Porter, R. F., Christensen, S. and Schiermacker-Hansen, P. (1996). *Field guide to the birds of the Middle East*. T. & A. D. Poyser, London.

THE PLATES

Plate 1. OVERALL SUMMARY OF FLYING RAPTORS

Every regularly occurring raptor in the Western Palearctic will resemble somewhat one of the 16 raptors shown here in flight from below. Similar raptors are noted, with numbers of plate where each is depicted given in brackets.

1. Griffon Vulture. Adult [14]. Cinereous Vulture [15].

2. White-tailed Eagle. Juvenile [10, 11]. Immature Eastern Imperial Eagle [34] and Immature Golden Eagle [35]. See also Plate 2.

3. Steppe Eagle. Adult [31, 32]. Eastern Imperial Eagle [34] and Golden Eagle [35]. See also Plate 2.

4. Osprey. Adult [5]. Distinctive, no other raptor similar.

5. Egyptian Vulture. Juvenile [13]. Immature Bearded Vulture [12].

6. Common Buzzard. Adult [24–27]. Rough-legged Buzzard [24–26], Long-legged Buzzard [24–26], Steppe Buzzard [24–26], and Eurasian Honey Buzzard [6, 7]. See also Plate 3.

7. Steppe Buzzard. Adult dark morph [27]. Dark-morph Long-legged Buzzard [27] and dark-morph Eurasian Honey Buzzard [7]. See also Plate 3.

8. Greater Spotted Eagle. Adult [29, 30]. Lesser Spotted Eagle [28, 30] and dark-morph Booted Eagle [37]. See also Plate 2.

9. Black Kite. Adult [8, 9]. Red Kite [8], dark-morph Booted Eagle [37], and juvenile Marsh Harrier [17].

10. Hen Harrier. Adult female [19]. Adult female Montagu's Harrier [21] and Pallid Harrier [21]. See also Plate 3.

11. Montagu's Harrier. Adult male [20]. Adult male Pallid Harrier [20], Hen Harrier [19], Marsh Harrier [16], and Black-shouldered Kite [20]. See also Plate 3.

12. Bonelli's Eagle. Juvenile [36]. Short-toed Snake Eagle [5], immature Eastern Imperial Eagle [34], and Booted Eagle [37]. See also Plate 2.

13. Peregrine. Juvenile [46]. Lanner Falcon [44], Saker Falcon [45], Gyrfalcon [45], Barbary Falcon [46], and Eleonora's Falcon [41]. See also Plate 4.

14. Common Kestrel. Adult female [38, 39]. Lesser Kestrel [38, 39], Merlin [43], and Red-footed Falcon [40]. See also Plate 4.

15. Eurasian Sparrowhawk. Adult female [23]. Northern Goshawk [23] and Levant Sparrowhawk [22].

16. Red-footed Falcon. Adult male [40]. Eurasian Hobby [42] and dark-morph Eleonora's Falcon [41]. See also Plate 4.

Plate 1

Plate 2. SUMMARY OF FLYING EAGLES

Eagles are large to medium-sized raptors. Most species (*Aquila* and *Hieraaetus*) have completely feathered tarsi, but sea and snake eagles have bare tarsi.

Every regularly occurring eagle in the Western Palearctic is shown here in flight from below; most are shown in adult and juvenile plumages, if different.

1. **White-tailed Eagle.** Juvenile [11]. Older immatures [10, 11] distinguishable.

2. **White-tailed Eagle.** Adult [10].

3. **Golden Eagle.** Juvenile [35]. Older immatures distinguishable.

4. **Golden Eagle.** Adult [35].

5. **Short-toed Snake Eagle.** Adult [5]. Juvenile barely distinguishable.

6. **Eastern Imperial Eagle.** Juvenile [34]. Second plumage similar.

7. **Eastern Imperial Eagle.** Adult [34]. Spanish Imperial Eagle [33] similar.

8. **Bonelli's Eagle.** Juvenile [36]. Older immatures streaked.

9. **Bonelli's Eagle.** Adult [36].

10. **Steppe Eagle.** Juvenile [32]. Older immatures similar.

11. **Steppe Eagle.** Adult [31].

12. **Booted Eagle.** Light-morph [37]. Juvenile identical.

13. **Booted Eagle.** Rufous-morph [37]. Juvenile identical.

14. **Booted Eagle.** Dark-morph [37]. Juvenile identical.

15. **Lesser Spotted Eagle.** Adult [28, 30]. Immatures similar.

16. **Greater Spotted Eagle.** Adult [29, 30]. Immatures similar.

Plate 2

Plate 3. SUMMARY OF FLYING HARRIERS AND BUZZARDS

Harriers are long-winged, long-tailed raptors that hunt in a distinctive slow quartering flight low over the ground. Four species of harriers occur regularly throughout the Western Palearctic.

1. **Marsh Harrier.** Adult female [17]. Juvenile is similar.

2. **Marsh Harrier.** Adult male [16].

3. **Montagu's Harrier.** Juvenile [21]. Pallid Harrier juvenile is similar.

4. **Montagu's Harrier.** Adult female [21].

5. **Montagu's Harrier.** Adult male [20].

6. **Pallid Harrier.** Adult male [20].

7. **Pallid Harrier.** Adult female [21].

8. **Hen Harrier.** Adult male [19].

9. **Hen Harrier.** Adult female [19]. Juvenile is similar.

Buzzards are medium-sized raptors with robust bodies, small beak, long, broad wings, and medium-length tail. Honey Buzzards are actually specialized kites but appear similar to buzzards, particularly in flight. One species of Honey Buzzard and three species (four forms) of buzzards occur throughout the Western Palearctic.

10. **Eurasian Honey Buzzard.** Adult male [6, 7].

11. **Eurasian Honey Buzzard.** Adult female [6, 7]. Juvenile is somewhat similar.

12. **Long-legged Buzzard.** Adult [24–26].

13. **Long-legged Buzzard.** Dark-morph adult [27].

14. **Rough-legged Buzzard.** Adult female [24–26]. Adult male and juvenile similar.

15. **Common Buzzard.** Adult [24–27]. Juvenile is usually similar.

16. **Steppe Buzzard.** Dark-morph adult [27]. Juvenile is similar.

17. **Steppe Buzzard.** Rufous adult [24–26]. Juvenile is usually somewhat different.

Plate 3

Plate 4. SUMMARY OF FLYING FALCONS

Falcons are small- to medium-sized raptors that have long pointed wings, dark eyes, and notched beaks.

1. **Merlin.** Adult male [43].
2. **Merlin.** Adult female [43]. Juveniles similar.
3. **Common Kestrel.** Adult female [39]. Juveniles similar.
4. **Common Kestrel.** Adult male [38].
5. **Lesser Kestrel.** Adult female. [39]. Juveniles similar.
6. **Lesser Kestrel.** Adult male [38].
7. **Red-footed Falcon.** Juvenile [40].
8. **Red-footed Falcon.** Adult female [40].
9. **Red-footed Falcon.** Adult male [40].
10. **Eurasian Hobby.** Juvenile [42].
11. **Eurasian Hobby.** Adult [42].
12. **Sooty Falcon.** Juvenile [42].
13. **Sooty Falcon.** Adult [42].
14. **Eleonora's Falcon.** Juvenile [41].
15. **Eleonora's Falcon.** Dark-morph adult [41].
16. **Eleonora's Falcon.** Adult [41].
17. **Barbary Falcon.** Juvenile [46].
18. **Barbary Falcon.** Adult [46].
19. **Lanner Falcon.** Adult [44].
20. **Lanner Falcon.** Juvenile [44].
21. **Peregrine.** Juvenile [46].
22. **Peregrine.** Adult [46].
23. **Gyrfalcon.** Juvenile [45].
24. **Gyrfalcon.** Adult [45].
25. **Saker Falcon.** Juvenile [45].
26. **Saker Falcon.** Adult [45].

Plate 4

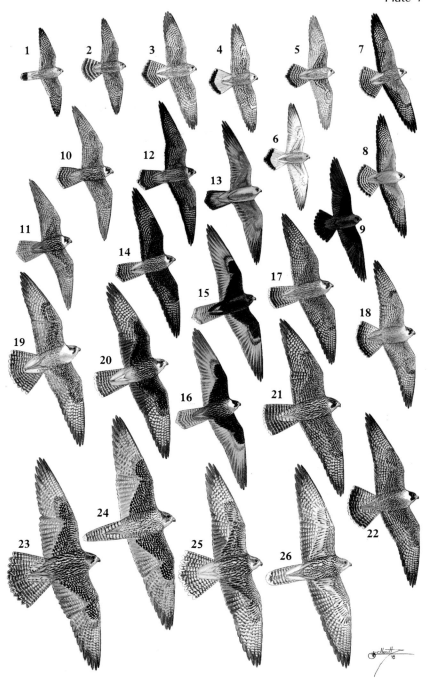

Plate 5. SHORT-TOED SNAKE EAGLE AND OSPREY

1. Short-toed Snake Eagle *Circaetus gallicus* p. 84; photos, p. 308

Medium-sized pale eagles; only large eagle that hovers. Plumages of adults and juveniles are almost alike. Sexes are almost alike in plumage. Wingtips reach tail tip on perched eagles.

[a] Adult. Whitish below except for dark head and neck, with bold barring on belly and underwings. Males tend to have whitish vertical streaks on the breast.

[b] Adult head. Females tend to have unmarked dark brown throat and upper breast.

[c] Juvenile trailing edge of underwing. Lacks outer dark band through dusky band on trailing edge.

[d] Adult secondary, With outer dark band in subterminal band.

[e] Juvenile secondary, lacking outer dark band in subterminal band.

[f] Adult. Pale headed variety (see also 1[j]). Note large pale patches on coverts.

[g] Juvenile. Breast is rufous-brown, and eye is lemon yellow.

[h] Adult. Shows large head, orange-yellow eye, and dark brown breast. Wingtips reach tail tip.

[i] Silhouette. Glides with wings held flat, with tips drooped, or in a modified dihedral.

[j] Some individuals are extremely pale, with pale head and lacking dark hood and bold underwing barring (see also 1[f]).

[k] Hovering with tail depressed and fanned, often with legs dangling.

Similar species: Osprey, Bonelli's Eagle, Eurasian Honey Buzzard

2. Osprey *Pandion haliaetus* p. 14; photos, p. 288

Large, pale, long-winged and long-legged raptors, like medium-sized eagles. Their *gull-like crooked wings* are distinctive. Sexes are almost alike in plumage, but females average somewhat larger than males. Wingtips extend just beyond tail tip on perched birds.

[a] Adult. Note *dark eye stripe.*

[b] Adult. Note white body and dark carpal patches.

[c] Juvenile. Note pale feather edges on back and upperwing coverts, dark streaks in crown, and orangish eye.

[d] Adult. Females usually have more boldly marked necklaces, but sexes overlap completely in this character.

[e] Distant flying bird showing orangish cast to uppertail.

[f] Distant bird soaring with straight wings.

[g] Silhouette. Showing gull-like wing attitude.

Similar species: Short-toed Snake Eagle, Bonelli's Eagle

Plate 5

1[d] 1[e]

1[g] 1[h]

1[b]

1[a]

1[f]

1[c]

1[i]

1[j]

1[k]

2[f]

2[g]

2[b]

2[e]

2[c]

2[a]

2[d]

Plate 6. PALER HONEY BUZZARDS

1. Eurasian Honey Buzzard *Pernis apivorus* p. 20; photos, p. 290

Large, buzzard-like kite. All but palest individuals show an oval or rectangular dark carpal patch on each underwing. Coloration of plumages of adult males, adult females, and juveniles are almost alike, but eye, cere, and face colour, amount of dark on wingtips, pattern of marking on underparts, and tail and secondary pattern distinguish them.

[a] Adult female. Pale-bodied individual. Adults have dark ceres and yellow eyes. Underparts are marked with spots and barring. Note small dark malar stripe and dark barring across secondaries of adult females.

[b] Flight silhouette. Note drooped wingtips.

[c] Adult male. Males have grey faces, and adults have dark ceres and yellow eyes.

[d] Juvenile. Juveniles have yellow ceres, brown eyes, and streaked underparts. Wingtips fall somewhat short of tail tip.

[e] Adult male. Palest variant. Only tips of outer flight primaries are black, contrasting with rest of white primaries. Males have grey faces and wide unbarred area on secondaries. Pattern and colour of secondaries are same as those of primaries.

[f] Adult female. Typical female with tips of outer primaries black but dusky to emarginations, lacking sharp line of contrast as on those of males.

[g] Juvenile. Palest juvenile with whitish head and lack of large dark carpal patches. Dark secondaries of juveniles contrast with paler primaries and coverts.

[h] Juvenile. Typical juvenile with yellow cere, brown eyes, streaked underparts, and wide dark tips on outer primaries.

[i] Juvenile. More heavily marked individual.

[j] Juvenile. Note pale primary patches, pale 'U' of uppertail coverts, and tail pattern. Juveniles often show a narrow pale bar across middle of upperwings (tips of greater wing coverts).

[k] Adult male. More heavily marked male.

[l] Adult female. Note tail pattern, dark cere, and yellow eye. Adult females show pale primary patches on upperwings.

Similar species: Common Buzzard, Black Kite

2. Crested Honey Buzzard *Pernis ptilorynchus* p. 25

Vagrant. Similar to Eurasian Honey Buzzard but with wider wings and six 'fingers' on wingtips. They lack dark carpal patches on underwings.

[a] Juvenile. Much like Eurasian Honey Buzzard juvenile but with wider wings and six 'fingers' on wingtips. Also lacks dark carpal patches on underwings.

Similar species: Eurasian Honey Buzzard

Plate 6

Plate 7. DARKER HONEY BUZZARDS

1. Eurasian Honey Buzzard *Pernis apivorus* p. 20, photos, p. 290

Large, buzzard-like kite. All but palest individuals show an oval to rectangular dark carpal patch on each underwing. Coloration of plumages of adult males, adult females, and juveniles are almost alike, but eye, cere, and face colour, amount of dark on wingtips, and tail and secondary pattern distinguish them.

[a] Adult male. Males have grey faces. Note short legs and horizontal posture when perched on ground. Adults have yellow eyes and dark ceres.

[b] Adult female. Dark morph. Note tips of outer primaries: dark on tips and dusky down to emargination, lacking sharp line of contrast as on those of males. Most adult females show barring across secondaries.

[c] Adult female. Heavily marked individual. Wingtips do not reach tail tip. Posture is more upright when perched on tree limb. Some adult females show dark eye-lines.

[d] Juvenile. Juveniles have yellow ceres, brown eyes, and streaked underparts.

[e] Adult male. Dark morph. Only tips of outer flight primaries are black; these contrast with rest of white feathers. Males have grey faces.

[f] Juvenile. Dark morph. Note yellow cere, brown eye, and wide dark tips to outer primaries. Note pale buffy bar across underwings of juveniles.

[g] Adult male. Rufous morph. Only tips of outer primaries are black on males. Note grey face and wide unbarred area across secondaries of adult males.

[h] Adult female. Heavily marked individual. Note short dark malar stripes of females. This individual shows somewhat male-like pattern on secondaries.

[i] Adult male. Heavily marked individual. Only tips of outer primaries are black on males. Note grey face and wide unbarred area across secondaries of adult males.

[j] Juvenile. Juveniles have yellow ceres, brown eyes, and wide dark tips to outer primaries. Note pale buffy bar across underwings of juveniles.

[k] Adult male. Uppersides of males are brownish-grey. Note large dark commas at wrists and grey face.

Similar species: Common Buzzard, Black Kite

2. Crested Honey Buzzard *Pernis ptilorynchus* p. 25

Vagrant. Similar to Eurasian Honey Buzzard but with wider wings and six 'fingers' on wingtips. They lack dark carpal patches on underwings. Their tail patterns are similar.

[a] Adult female. Dark morph. Like Eurasian Honey Buzzard but with wider wings and six 'fingers' on wingtips.

[b] Adult male. Males have dark brown eyes and single wide pale tail band. Note lack of dark carpal patches.

[c] Adult male. Like Eurasian Honey Buzzard but with wider wings and six 'fingers' on wingtips, dark brown eyes, and single wide pale tail band. Males have grey faces.

Similar species: Eurasian Honey Buzzard

Plate 7

Plate 8. RED AND BLACK KITES

1. **Red Kite** *Milvus milvus* p. 33; photos, p. 294

The Western Palearctic's largest kite. Compared to Black Kites, they are larger, more slender, overall more rufous, and have longer, more deeply forked tails. *Large squarish white primary panels on underwings are diagnostic. Wingtips barely reach tip of shortest (central) tail feathers on perched birds, but far from tail tip.* Sexes are alike in plumage; females are noticeably larger. Juvenile plumage is different from that of adult.

[a] Adult. Head is completely white, with orange-yellow cere and base of beak. Shoulders and underparts are rufous. Wingtips fall way short of tip of long rufous tail.

[b] Juvenile. Head is pale buff, with dark eye and all-dark beak. Buffy-rufous underparts have pale streaks. Tail shorter and less deeply forked than that of adult.

[c] Adult. Rufous underparts, large square white primary panels, and long, deeply forked tail are diagnostic. Note narrow black band through underwings.

[d] Juvenile. Underparts are buffy-rufous; with belly noticeably paler. Note large square white primary panels and wide black band and narrow pale line through underwings. Tail is less deeply forked than that of adult.

[e] Adult. In flight, white head, chestnut lesser coverts, and long deeply forked tail are diagnostic. Note orange-yellow cere and base of beak.

[f] Juvenile. Like adult but tail less deeply forked, with dark subterminal band. Note narrow pale band across upperwings.

[g] Head-on silhouette. Wings bowed: wrists up and tips down.

Similar species: Black Kite

2. **Black Kite** *Milvus migrans* p. 37; photos, p. 296

Typical adult perched and adult and juvenile in flight are shown. Variations are shown on Plate 9.

Black Kites are overall dark brown, often with a reddish cast on underparts of adults. Most birds show pale crescent-shaped to squarish primary panel on otherwise dark underwing; dark barring in panels noticeable on close birds. *Wingtips reach beyond central tail feathers to near tail tip on perched birds.* Tail is forked with outer tail feathers somewhat longer than central ones but appears square when fully fanned; occasional birds (usually juveniles) have rounded tail tips, lacking any fork. Sexes are alike in plumage; females are noticeably larger. Juvenile plumage is different from that of adult. Cere and legs are yellow. Beak is all black.

[a] Adult. Head has whitish face and crown, dark ear patches, brownish nape, and all-dark beak. Underparts are dark sooty-red. Wingtips almost reach tip of brown tail.

[b] Adult. Underparts are dark sooty-red. Primary panels are smaller and less well defined than those of Red Kites. Tail has shallow fork.

[c] Juvenile. Buffy head shows dark ear patches. Underparts have pale streaks. Pale primary panels are smaller than those of Red Kites.

Similar species: Red Kite, juvenile Marsh Harrier, dark-morph Booted Eagle

Plate 8

Plate 9. BLACK KITES

Typical adult perched and adult and juvenile in flight from below shown on Plate 8. Plumage variations are shown on this plate.

Black Kite *Milvus migrans*   p. 37; photos, p. 296

Black Kites are overall dark brown, often with a reddish cast on underparts of adults. Most birds show whitish crescent-shaped to squarish primary panel on each otherwise dark underwing; dark barring in panels noticeable on close birds. *Wingtips reach beyond central tail feathers to near tail tip on perched birds.* Tail is forked with outer tail feathers somewhat longer than central ones but appears square when fully fanned; occasional birds (usually juveniles) have rounded tail tips, lacking any fork. Sexes are alike in plumage; females are noticeably larger. Juvenile plumage is different from that of adult. Cere and legs are yellow. Beak is all black.

[a]   Adult. Pale bars on upperwings and brownish tail with shallow fork.

[b]   Dark extreme juvenile. Overall dark sooty brown.

[c]   Juvenile. Head with dark ear patches. Underparts streaked whitish. Wingtips almost reach tail tip.

[d]   Head of Egyptian adult. Yellow beak.

[e]   Juvenile. Pale tips on back and upperwing coverts give scaly appearance. Pale tips of greater coverts form a narrow pale line through mid-wing.

[f]   Dark extreme adult. All dark sooty brown. No primary panels or tail banding.

[g]   Reddish adult. Seen occasionally in Middle East. May be hybrid with Red Kite.

[h]   *lineatus* juvenile. Underparts heavily streaked.

[i]   Darker juvenile. Occasionally tail appears more rounded and not forked. Tips of greater coverts form pale band through midwing. Small whitish crescent-shaped primary panels.

[j]   Silhouette. Bowed wings: wrists up and tips down.

[k]   Adult fishing.

[l]   Adult. Rapid glide silhouette. Note crescent-shaped pale primary panels.

[m]   Pale extreme juvenile. Often show large whitish underwing panels.

Similar species: Red Kite, juvenile Marsh Harrier, dark-morph Booted Eagle

Plate 9

[a] [b] [c] [d] [e] [f] [g] [h] [i] [j] [k] [l] [m]

Plate 10. WHITE-TAILED EAGLE

White-tailed Eagle *Haliaeetus albicilla*   p. 47; photos, p. 298

Large, dark eagles that have bare tarsi, long broad wings with straight leading and trailing edges, and short, rather wedge-shaped tails. Adult plumage is attained in 4 or 5 years. Wingtips reach tail tip on perched eagles.

[a] Adult. Adults are mostly brown with pale head and neck and white tail. Head and neck protrude from body almost as much as tail.

[b] Second winter. Similar to juvenile (shown on Plate 11) but with whitish mottling on underparts and yellow cere. Note ragged trailing edge of wings due to moult.

[c] Third winter. Similar to juvenile and second winter eagles but underparts darker. Note smooth trailing edge of wings. Often one or more longer, more pointed juvenile secondaries are retained, as shown on this eagle.

[d] Fourth winter. Three-year-old plus eagles are more adult-like, with pale beak and eyes but still show darker head and neck, black markings on tail, and dusky areas on beak.

[e] Adult. Adults are mostly brown with pale head and neck and white tail. Upperparts are rather two-toned: paler back and upperwing coverts contrast with darker flight feathers. Often show black spots on uppertail coverts, as shown here. Trailing edges of inner wings can appear a bit bowed, depending on how wings are held.

[f] Second winter. Similar to juvenile (shown on Plate 11) but with whitish mottling on back and upperwing coverts and yellow cere. Note ragged, uneven trailing edge of wings due to moult.

[g] Fourth winter. Three-year-old plus eagles are more adult-like, with pale beak and eyes but show darker head and neck, retained immature tail feathers, black markings on white tail feathers, and dusky areas on beak.

[h] Fifth winter. Some first plumage adults have darker heads and upper breasts and often show some dark tips to tail feathers.

[i] Third winter. Similar to second winter eagles but eye is paler and cere and beak are yellow, with a dusky area on tip of latter.

[j] Fourth winter. Three-year-old plus eagles are more adult-like, with pale beak and eyes but still show darker head, black markings on tail, and dusky areas on beak.

[k] Adult. Adults are mostly brown with pale head and neck and white tail. Note large lemon-yellow beak, yellow eyes, and pale back and upperwing coverts.

Similar species: Eastern Imperial Eagle, Greater Spotted Eagle

Plate 10

[a]

[b]

[c]

[d]

[e]

[f]

[g]

[h]

[i]

[j]

[k]

Plate 11. IMMATURE SEA EAGLES

1. White-tailed Eagle *Haliaeetus albicilla*   p. 47; photos, p. 298

Juvenile and second winter White-tailed Eagles appear overall rather mottled. Wingtips usually show seven 'fingers', and tail tip shows white spikes.

[a] Juvenile. Uppersides are somewhat two-toned; back and upperwing coverts are mottled tawny and brown and appear paler than flight feathers.

[b] Juvenile. Breast and belly are same colour. Axillaries are mostly white. Note narrow white band through underwing.

[c] Juvenile. Beak, cere, and eyes are dark. Wingtips fall a bit short of tail tip. Secondaries are all same length and colour.

[d] Second winter. Appears overall more mottled with white compared to juvenile. Eyes and cere are pale. Leg feathers are uniformly dark. Wingtips almost reach tail tip. Secondaries are a mix of new shorter, darker and old longer, paler feathers.

Similar species: Eastern Imperial Eagle, Greater Spotted Eagle

2. Bald Eagle *Haliaeetus leucocephalus*   p. 52

Vagrant to the British Isles from North America. Juveniles and second winter Bald Eagles are overall dark brown, with some whitish areas. Wingtips usually show six 'fingers', and tail tip shows even dark band.

[a] Juvenile. Uppersides are somewhat two-toned; back and upperwing coverts are uniform tawny-brown and appear paler than flight feathers.

[b] Second winter. White belly contrasts with dark breast. Note wide pale superciliaries and longer retained juvenile secondaries.

[c] Juvenile. Tawny belly contrasts with dark breast. Axillaries are mostly white. Note narrow white band through underwing.

[d] Juvenile. Overall darker than juvenile White-tailed Eagle. Wingtips fall a bit short of tail tip.

Similar species: White-tailed Eagle

3. Pallas's Fish Eagle *Haliaeetus leucoryphus*   p. 44

Vagrant to northern Europe, Middle East, and the Caucasus from central Asia. Juveniles and second winter Pallas's Fishing Eagles are overall pale brown, with completely dark tail. Wings are narrower than two above species.

[a] Juvenile. Upperparts are contrastingly two-toned; back and upperwing coverts are much paler than flight feathers.

[b] Juvenile. Underparts are pale brown. Note white band through underwings; it and white inner primaries form a pale 'M' on underwings, noticeable even on distant eagles. New darker head feathers of first winter Eagles give a 'dark hooded' look.

[c] Juvenile. Overall pale brown with dark ear patches. Wingtips fall a bit short of tail tip.

Similar species: White-tailed Eagle, juvenile Bonelli's Eagle

Plate 11

Plate 12. BEARDED VULTURE

Bearded Vulture *Gypaetus barbatus* p. 56; photos, p. 302

Distinctive large raptor with long wedge-shaped tail and bristle of dark feathers extending from lower mandible. Pale head and underparts are usually rubbed with iron oxide and, as a result, have a rufous-buff wash.

[a] Juvenile. Note dark head, whitish triangle on back, and greyish-brown underparts. Beard is short. Second winter birds are essentially like juveniles.

[b] Fourth winter plus (43–60 months old). Head whitish with fine dark streaking. Pale underparts usually show rufous-buff wash. Beard is fairly long.

[c] Adult. Note drooped wings and shape of tail.

[d] Adult. Head and underparts are pale rufous-buff; this from iron oxide in soils. Note red eye-ring and black mask across face. Beard is fairly long. Upperparts are steely-grey with narrow white shaft streaks. Note narrow dark band across breast.

[e] Juvenile. Wings are broader and tail less wedge-shaped compared to those of adults. Note blackish head and greyish-brown underparts.

[f] Third winter plus (21–36 months old). Similar to juveniles but with a variable amount of shorter, darker secondaries and rufous-buff wash on pale underparts. Old juvenile secondaries form ragged trailing edge of wings.

[g] Fourth winter plus (43–60 months old). Similar to adults but with thicker band across upper breast, darker axillaries, and a few retained juvenile secondaries.

[h] Adult. Shows classic shape. Note whitish axillaries and smooth trailing edge of wings.

[i] Juvenile. Note dark head and upperparts with white triangle on back. Wings are broader and tail is less deeply wedge-shaped that those of adults.

[j] Third winter plus (21–36 months old). Similar to juvenile but lacking white triangle on back. Note paler head with dark eyebrow and new shorter secondaries.

[k] Adult. Uppersides are steely-grey with narrow white shaft streaks.

[l] Adult. Showing bone-dropping behaviour.

Similar species: Egyptian Vulture

Plate 12

Plate 13. EGYPTIAN VULTURE

Egyptian Vulture *Neophron percnopterus*   p. 61; photos, p. 300

Small vulture with long thin beak. In flight *wedge-shaped tail* distinctive. There are five annual plumages, progressing from juvenile to adult.

[a]   Adult. Black and white plumage. Bare face skin varies from yellow to orangish-yellow.

[b]   Adult head. Blackish smudge under eye is not a sexual character.

[c]   Juvenile. Overall darkish brown, with bluish face. Dark forehead feathers form a 'widow's peak'.

[d]   Second winter. Like juvenile but tail coverts are same colour as body, whitish down on crown, and forehead lacks the 'widow's peak'.

[e]   Third winter. Variable amount of whitish mottling overall; face is yellowish.

[f]   Fourth winter. Almost like adult, but a few dark feathers throughout, especially dark neck.

[g]   Fourth winter. Almost like adult, but a few dark feathers throughout, especially dark neck. Darker bird, especially neck.

[h]   Juvenile. Overall dark except for whitish undertail coverts.

[i,j]   Juveniles. Variable amount of whitish on upperparts from none (i) to lots (j), and whitish uppertail coverts.

[k]   Second winter. Like juvenile but uppertail coverts are brownish and uppersides of flight feathers show a greyish cast. Crown is now whitish.

[l]   Second winter. Like juvenile but undertail coverts are brownish.

Third winter plumage is highly variable; whitish feathers replace brownish ones over 2 years.

[m]   Third winter. Overall mottled whitish.

[n]   Third winter. Note wide white bars on median coverts; greater coverts are brown.

[o,p]   Fourth winter birds. Almost like adults, but some dark feathers retained, especially on neck, forming a collar. Greater underwing coverts are mostly white. Some new tail feathers are white. May or may not have black patch on inner greater upperwing secondary coverts.

[q,r]   Adults. Overall black and white, with rusty soiling on neck, parts of upperwings, and sometimes on belly.

[s]   Silhouette. Wingtips are somewhat below wrists during glides.

Similar species: Light-morph Booted Eagle, juvenile Bearded Vulture

Plate 13

Plate 14. GRIFFON AND RÜPPELL'S VULTURES

1. Griffon Vulture *Gyps fulvus*   p. 70; photos, p. 304

Large tawny-brown to pale brown vulture. Often shows two circular bare areas on sides of upper breast. All age classes appear rather similar.

[a] Adult. Have yellow to yellow-brown eyes, pale yellow beaks, and short white ruff at base of neck. Note full crop.

[b] Older immature. Similar to adult but with dark eyes, partially pale beak, mottled upperwing coverts, and rather unruly white ruff.

[c] Juvenile. Somewhat similar to adult but with dark eyes, dark beak, long lancelot brown ruff feathers, and noticeable white streaking on underparts and upperwing coverts. Note two circular bare areas, which can vary from blue to red.

[d] Adult. Typical *Gyps* vulture overall shape. Body and coverts are tawny-brown to brown with short narrow white bar in centre and long wide pale bar across greater coverts on each underwing. Note pale secondaries with dark tips.

[e] Juvenile. Similar to adult but with more noticeable white streaking and uniformly dark and pointed-tipped secondaries.

[f] Adult. Tawny-brown to brown back and upperwing coverts contrast with dark brown flight and tail feathers.

[g] Adult. When landing, Griffon Vulture's head and neck are extended and lowered. Note white patches on inner legs and bare patch at base of hind-neck.

[h] Flight silhouette. Griffon Vultures soar with their wings in a dihedral.

[i] Flight silhouette. They glide with wings level or with wingtips slightly drooped.

Similar species: Cinereous Vulture

2. Rüppell's Vulture *Gyps rueppellii*   p. 66

Large dark African vulture, vagrant to north Africa. Juveniles appear darker than do adults.

[a] Adult. Adults have pale eyes, orangish-yellow beak, short white ruff at base of neck, and creamy to whitish edges on body and covert feathers.

[b] Juvenile. Overall dark brown, except for white head and neck and faint buffy shaft streaks and narrow feather edges on most body and covert feathers. Eyes and beak are dark, and ruff is composed of long brown lanceolate feathers.

[c] Adult. Typical *Gyps* vulture overall shape. Body appears quite pale due to wide creamy feather edges. Dark underwings show a narrow white bar in lesser coverts and some creamy scalloping. Paler secondaries with dark tips can be seen in good light.

[d] Juvenile. Typical *Gyps* vulture overall shape. Overall dark brown with faint buffy shaft streaks on body and white bar on lesser wing coverts of each wing.

[e] Adult. Creamy scalloped back and upperwing coverts contrast with dark brown flight and tail feathers. Upperwing coverts often appear completely whitish.

Similar species: Griffon and Cinereous Vultures

Plate 14

Plate 15. LARGE DARK VULTURES

1. Cinereous Vulture *Aegypius monachus*   p. 74; photos, p. 306

Cinereous Vultures are large dark, eagle-like vultures. Flying birds lack the secondary bulge of Griffon Vultures; front and rear edges of wings are nearly parallel. Show serrated trailing edge of wings in all plumages. Tails of flying birds are wedge-shaped. Pale legs and feet are noticeable on close birds. Adults and juveniles are almost alike; juveniles are overall darker.

[a] Adult. Pale head of adults contrasts with black throat and foreneck. Breast shows buffy streaking. Cere is usually pale bluish. Eyes are reddish-brown.

[b] Older immature. Head is similar to that of adult, but face skin is pale and crown and sides of neck are darker.

[c] Juvenile. Appears blacker overall, especially head. Cere and face skin are usually pinkish but can change to pale blue in time. Eyes are dark brown.

[d] Adult. Appears more brownish-black, with pale head and faint narrow pale line along base of flight feathers.

[e] Juvenile. Appears blackish overall.

[f] Adult. Note uniform colour on upperwings. Appears paler above than juveniles (not shown). Some adults (males?) show small white patches on upperwing coverts.

[g] Display of adult showing 'brushes'. Brushes of adult are paler than colour of back; those of non-adults are same colour as back.

[h] Display of adult showing 'brushes'.

[i] Silhouette. Glides with wings drooped.

Similar species: Lappet-faced Vulture, White-tailed Eagle, Griffon Vulture

2. Lappet-Faced Vulture *Torgos tracheliotus*   p. 78; photos, p. 302

Nominate race occurs locally in Morocco and is shown on Plate 47.

Lappet-faced Vultures are an African species; race *negevensis* occurs in Israel (formerly), Jordan, and Sinai. It is similar in size and plumage to Cinereous Vulture, but *undertail coverts are pale and undersides and underwings show whitish areas.* Note: thighs of adults are dark, different from African races. Lappets on side of neck are small and indistinct, unlike those of African birds. Adults and juveniles are similar in plumage. Iris is dark brown.

[a] Adult. Middle Eastern birds have dark leg feathers and pinkish heads.

[b] Juvenile. Similar to adult but overall darker. Head is covered with whitish down, forming a distinctive peak on rear crown.

[c] Adult. Note pale undertail coverts, pale lines on underwings, and short tail.

[d] Juvenile. Note white head, pale undertail coverts, serrated trailing edge of wings, and short tail.

[e] Adult. Shows two-toned upperwings. Upperwing pattern of juvenile (not shown) is the same.

[f] Silhouette. Glides on flat wings.

Similar species: Cinereous Vulture, White-tailed Eagle, Griffon Vulture

Plate 15

Plate 16. MARSH HARRIER MALES

Marsh Harrier *Circus aeruginosus*   p. 92; photos, p. 310

The largest and darkest harrier of the Western Palearctic; lacks white uppertail coverts in all plumages. Adult males and females have different plumages.

[a] Adult male. Facial ring is noticeable. Note dark brown back, grey on secondaries, and rufous leg feathers.

[b] Second winter male. Much like adult female, but with streaking on nape and upper breast, greyish cast on secondaries, and subterminal tail band.

[c] Adult male. Note *tri-coloured pattern of upperparts*.

[d] Adult male. Note *rufous belly and paler breast*. Underwing coverts are whitish, sometimes with light rufous wash on younger birds.

[e] Third winter male underwing. Like adult but with dark subterminal band on secondaries and darker rufous coloration on underwing coverts.

[f] Second winter male. Appears similar to adult female but distinguished by grey on secondaries and primary coverts and streaking on nape.

[g] Second winter male tail. Showing heavy banding of some individuals.

[h] Second winter male. Streaked upper breast and pale secondaries distinguish them from adult females.

Similar species: Adult males of other harriers lack brown and rufous on bodies.

Plate 16

Plate 17. MARSH HARRIER ADULT FEMALES AND JUVENILES

Marsh Harrier *Circus aeruginosus*   p. 92; photos, p. 310

Largest and darkest harrier of the Western Palearctic; lacks white uppertail coverts in all plumages.

Females are larger than males. Juvenile plumage of sexes is alike and similar to that of adult females.

[a] Adult female. Note *creamy breast band*, creamy patches on wing coverts, and rufous leg feathers.

[b] Juvenile female. Appears somewhat like adult female, but lacks rufous tones and often lacks creamy patches on coverts and breast that are present on this individual. Juvenile females have dark brown eyes, males' are grey-brown.

[c] Adult female. Note large pale patch in primaries and rufous cast to tail.

[d] Adult female underwing. Shows extensive creamy mottling on coverts.

[e] Adult female. Note creamy patches on back and wing coverts and rufous cast to tail. Secondaries are often darker than and contrast with coverts. Uppertail coverts are rufous.

[f] Adult female head. Note yellow eye and streaks on crown.

[g] Juvenile head. Shows variant with only creamy nape patch.

[h] Juvenile head. Shows usual pattern, with dark eye and unstreaked crown. Male has yellow eye by spring; female retains brown eye for a year or two. Some juveniles show a diffuse crown patch.

[i] Juvenile. Note dark undersides of primaries with white crescent at base and brown body, coverts, and tail that lack rufous tones. Uppertail coverts are dark brown.

Similar species: Adult females and juveniles of other harriers show barring on undersides of flight feathers and banding in tails, and lack brown bodies.

Plate 17

Plate 18. DARK–MORPH MONTAGU'S AND MARSH HARRIERS

1. **Montagu's Harrier** *Circus pygargus* p. 103; photos, p. 314

Distinguished from dark-morph Marsh Harrier by its smaller size and narrow, rather pointed-tip wings. Lacks white uppertail coverts in all plumages. Light-morph birds are shown on Plates 20 and 21.

[a] Adult male. Body and most coverts are dark sooty grey, and tail, secondaries, and primary coverts are a more silvery grey. Note lack of black band across secondaries. Darker individual.

[b] Juvenile male underwing. Coverts and secondaries are dark brown, and primaries are pale, lacking barring.

[c] Adult male. Body and underwing coverts are sooty grey, and tail and secondaries are more silvery grey. Lacks black band across secondaries. Paler individual.

[d] Adult female. Overall dark brown, with faint tail bands.

[e] Adult female. Dark brown with paler barred primaries and undertail.

[f] Juveniles of both sexes (not illustrated) are identical and similar to adult female but with dark undersides of secondaries and dark brown or grey brown eyes.

Similar species: Dark-morph Marsh Harriers

2. **Marsh Harrier** *Circus aeruginosus* p. 92; photos, p. 310

Larger than dark-morph Montagu's Harriers, with wider wings and more rounded wingtips. Light-morph birds are shown on Plates 16 and 17.

[a] Adult male. Blackish-brown overall with greyish cast to primary upperwing coverts, inner primaries, and outer secondaries. Tail as in light-morph adult male.

[b] Adult male. White patches on underwings restricted to bases of inner primaries and outer secondaries. Can show rufous markings on breast and underwing coverts.

[c] Second winter male underwing. Underwing pale panel more mottled.

[d] Adult female. Overall dark brown with reduced pale primary panels and reddish cast to undertail. Can show small pale nape patch.

[e] Juvenile (not illustrated). Similar to adult female but lacks rufous cast to tail and yellow eye and has smaller white crescents on underwings.

Similar species: Dark-morph Montagu's Harriers, Black Kites, Dark-morph Booted Eagles

Plate 18

1[a]

1[b]

1[c]

1[d]

1[e]

2[a]

2[b]

2[c]

2[d]

Plate 19. HEN HARRIER

Hen Harrier *Circus cyaneus*   p. 98; photos, p. 312

Medium-sized harrier with rounded wingtips. On perched harriers, wingtips fall several cm short of tail tip.

[a] Adult female. Adult females show dark bands on secondaries and rufous blobs on leg feathers. Juveniles show narrow streaks on leg feathers.

[b] Adult male. *Grey hood and unmarked underparts.*

[c] Second winter male. Back darker than adult male's, with faint reddish streaking on underparts.

[d] Adult male. Distinguished by grey hood, large black patches on wingtips, and dark band on trailing edges of wings.

[e] Adult male. Note unmarked white uppertail coverts and rounded wingtips.

[f] Juvenile male head. Eye is pale grey-brown; females' eyes are dark brown. Cheek patches of juveniles lack buffy streaking of adult female's, thus appearing darker.

[g] Juvenile (American race *hudsonius*). Note rufous, mostly unstreaked underparts.

[h] Adult female. Adult females show two white bands across secondaries and rufous blobs on undertail coverts.

[i] Adult female. Note large unmarked white patch on uppertail coverts.

[j] Juvenile female. Eyes are dark brown, and cheek patches appear darker than those of adult female. Pale bands through secondaries are greyish, not whitish. Leg feathers and undertail coverts have rufous streaks, not blobs.

[k] Juvenile male underwing. Bands through secondaries are whiter than those of juvenile females.

[l] Silhouette. Flies with wings in a strong dihedral.

Similar species: Montagu's and Pallid Harriers

Plate 19

1. Black shouldered Kite *Elanus caeruleus*   p. 29; photos, p. 292

Small grey and white kite. Wingtips extend beyond tail tip.

[a] Adult *tail-cocking*.

[b] Adult. Red eye and black spots near eye. Wings often held below tail and tips extend beyond tail tip.

[c] Juvenile. Rufous wash on breast, brownish-grey back with white feather edges, grey crown, and orangish eye.

[d] Adult. Note all-black primaries and pointed wingtips.

[e] Juvenile underwing. Secondaries are greyish.

[f] Juvenile hovering. Back is brownish-grey; back and flight have white feather edges.

[g] Adult. Note black upperwing coverts.

Similar species: Adult male Pallid and Montagu's Harriers

2. Pallid Harrier *Circus macrourus*   p. 109; photos, p. 316

Small slender pointed-wingtip harrier. Adult female and juvenile are shown on Plate 21.

[a] Adult male. Body and underwing coverts white with black wedge on wingtips and little or no dark hood.

[b] Second winter male. Note faint hood, some rufous on outer tail feathers, a variable amount of reddish breast streaking, some black spotting on primaries and dusky tips of secondaries. Upperparts (not shown) are darker than those of adult males, often with a brownish cast.

[c] Adult male. Uppertail coverts do not appear as white patch.

Similar species: Adult male Montagu's and Hen Harriers, Black-shouldered Kites

3. Montagu's Harrier *Circus pygargus*   p. 103; photos, p. 314

Small slender pointed-wingtip harrier. Adult female and juvenile are shown on Plate 21. Dark-morph birds are shown on Plate 18.

[a] Adult male. Note grey upperparts with black line on uppersides of secondaries and rufous on outer tail feathers.

[b] Adult male. Note grey hood, rufous streaking on white belly, and black line through secondaries.

[c] Adult male. Variant with underparts entirely grey.

[d] Second winter male underwing. On some, primaries are not completely black.

[e] Gliding silhouette. Same as for Pallid Harrier. Strong dihedral.

Similar species: Adult male Pallid and Hen Harriers, Black-shouldered Kites

Plate 20

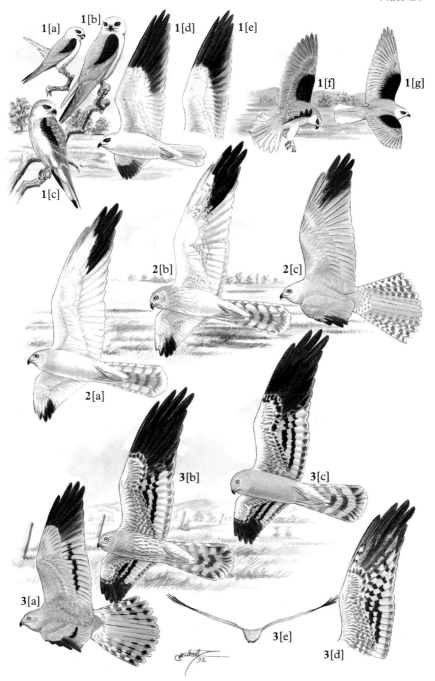

1[a]
1[b]
1[c]
1[d]
1[e]
1[f]
1[g]

2[a]
2[b]
2[c]

3[a]
3[b]
3[c]
3[d]
3[e]

Plate 21. ADULT FEMALE AND JUVENILE PALLID AND MONTAGU'S HARRIERS

1. Montagu's Harrier *Circus pygargus*   p. 103; photos, p. 314

Small slender harrier with pointed wingtips. *Wingtips reach tail tip on perched birds.*

[a] Juvenile female. *Lacks wide pale ring around head. Has some dark streaks on sides and upper breast.* Dark tips on inner primaries. Underparts and underwing coverts can be same colour as 2(a). Often shows barred axillaries.

[b] Juvenile male underwing. Primaries show more mottling, less barring than those of juvenile female.

[c] Adult female. *Wide white band through under secondaries extends to body.* Usually show barred axillaries. Wings are narrower at wrist than those of Pallid Harrier adult females.

[d] Adult female. *Facial ring is not distinct; it does not extend across throat.* Dark and pale bands on uppersides of secondaries (usually) obvious. Sides of slightly spread upper tail show three narrower pale bands.

[e] Adult female. Note shorter legs. *Wingtips reach tail tip on perched birds.* Facial ring does not extend across throat.

Similar species: Pallid and Hen Harriers

2. Pallid Harrier *Circus macrourus*   p. 109; photos, p. 316

Small slender harrier with pointed wingtips. *Wingtips fall a bit short of tail tip on perched birds.*

[a] Juvenile. *Wide pale ring around head. No streaking on sides.* Axillaries not boldly barred. Underparts and underwing coverts can be same colour as 1[a]. Tips of inner primaries usually lack dark tips.

[b] Adult female. Narrow white band through under secondaries does not extend to body. Secondaries appear darker than those of 1[c] and contrast with paler primaries. Tips of inner primaries usually lack dark tips. Axillaries not boldly barred. Wings are somewhat wider at the wrists compared to those of adult female Montagu's Harriers.

[c] Adult female. *Facial ring is diagnostic; facial ring extends across throat.* Uppersides of secondaries are uniformly dark. Sides of uppertail show two wide pale bands.

[d] Adult female. Note longer legs. *Wingtips fall a bit short of tail tip on perched birds.* Facial ring extends across throat.

Similar species: Montagu's and Hen Harriers

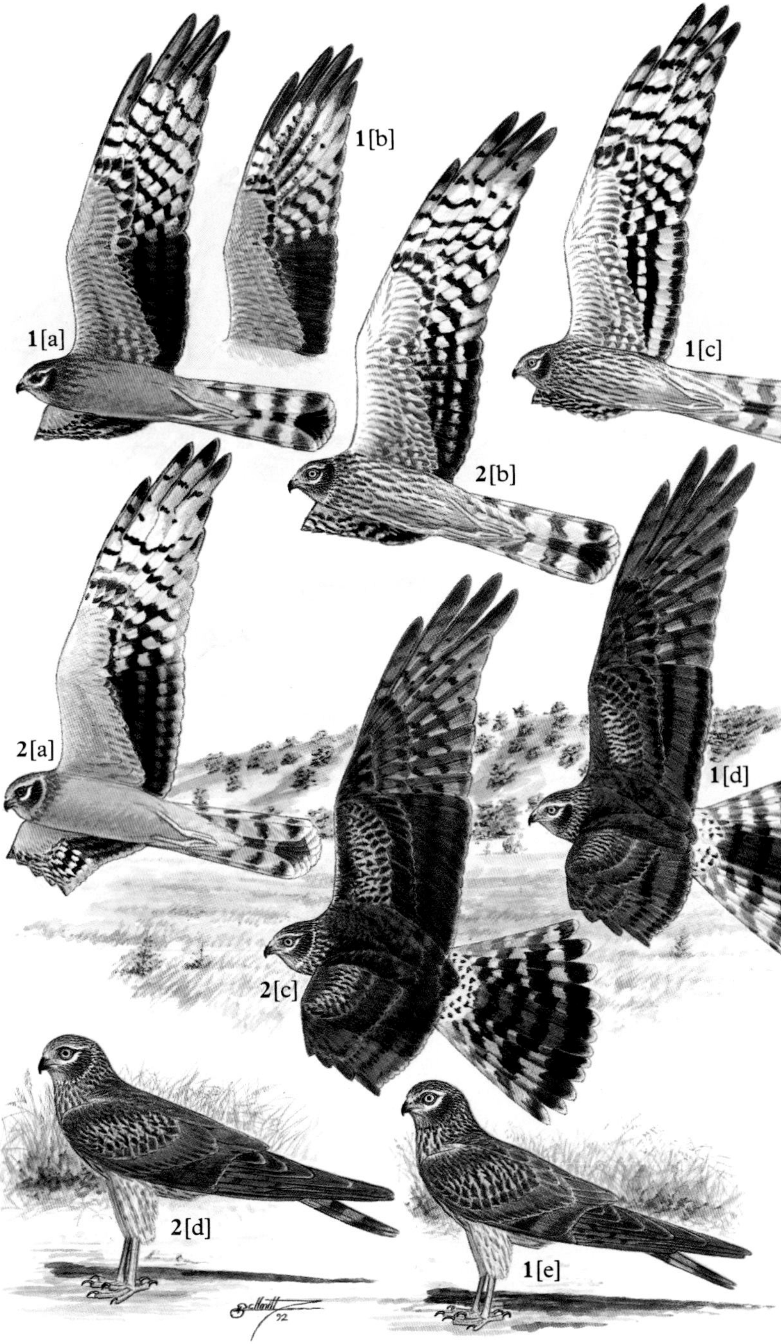
1[a]
1[b]
1[c]
2[b]
2[a]
1[d]
2[c]
2[d]
1[e]

Plate 22. LEVANT SPARROWHAWK AND SHIKRA

1. **Levant Sparrowhawk** *Accipiter brevipes*   p. 126; photos, p. 318

Small, pointed-wingtip hawk, with long primary projection when perched.

[a] Juvenile. Underparts are streaked with dark brown to rufous-brown spots forming wide streaks. Leg feathers are spotted. Note pale nape patch and brown eye.

[b] Adult female. Upperparts are blue-grey. Underparts are barred dark rufous-brown. Adult eye is dark brown, often with a reddish cast.

[c] Second summer female. Like adult female but with some retained body, covert, and tail feathers. Eye is dark brown. Spring into summer.

[d] Adult male. Upperparts are blue-grey. Underparts are barred pale rufous. Adult eye is dark brown, often with a reddish cast.

[e] Second summer male. Like adult male but with some retained body, covert, and tail feathers. Eye is dark brown. Spring into summer.

[f] Adult male. Unmarked underwings are pale with dark tips.

[g] Adult male. Upperparts are blue-grey, and spread uppertail shows faint banding.

[h] Adult female. Underparts and wing linings are heavily barred dark rufous-brown.

[i] Juvenile female. Underparts are streaked with dark spots, wing linings are heavily marked, and eye is brown. Autumn only.

[j] Second summer female. Like adult female but with retained juvenile flight feathers and tail and some body and covert feathers. Spring into summer.

[k] Juvenile male. Eye is brown, and tail shows numerous dark bands. Autumn only.

Similar species: Eurasian Sparrowhawk, Shikra, Lesser Kestrel

2. **Shikra** *Accipiter badius*   p. 122

Small, narrow-bodied hawk. Rare in Caucasus; vagrant to Middle East.

[a] Adult female. Upperparts are grey, underparts are barred rufous, and eye is orange-yellow.

[b] Adult male. Upperparts are grey, underparts are barred rufous, and eye is red.

[c] Juvenile. Underparts are streaked rufous-brown, with barred flanks. Leg feathers are spotted. Note shorter primary projection and pale yellow eye.

[d] Adult male. Upperparts are grey, and eye is red. Dark band shows on spread tail.

[e] Adult female. Wingtips appear rounded, and outer tail feathers are unbanded.

[f] Adult male. Wingtips can appear somewhat pointed in a glide. Note red eye, lightly banded flight feathers, and unbanded outer tail feathers.

[g] Juvenile female. Underparts are streaked rufous-brown, with barred flanks. Note relatively unmarked wing linings and pale yellow eye.

[h] Juvenile female. Upperparts are rufous-brown, and eye is pale yellow.

Similar species: Levant Sparrowhawk, Eurasian Sparrowhawk

3. **Eurasian Sparrowhawk** *Accipiter nisus*   p. 131; photos, p. 320

Covered in detail on Plate 23. Figures on this plate are for comparison.

[a,b] Juvenile female. Underparts are barred, eye is lemon-yellow, and supercilium is narrow and pale. Juveniles in fresh plumage show a narrow pale line across centre of upperwings.

Plate 22

Plate 23. NORTHERN GOSHAWK AND EURASIAN SPARROWHAWK

1. Northern Goshawk *Accipiter gentilis*   p. 137; photos, p. 322

Large, robust, relatively long-winged hawk.

[a] Adult male (northern). Note dark hood, wide whitish supercilium, robust body, barred underparts, fluffy white undertail coverts, and faintly banded tail.

[b] Adult male (southern). Like northern adult but darker overall.

[c] Juvenile. Note robust body with mottled brown back, wide pale supercilium, and long tail with wavy irregular banding, narrow white highlights, wide white terminal band, and wedge-shaped tip. Underparts are rufous-buff in fresh plumage.

[d] Adult. Overall grey above. Note somewhat pointed wingtips in powered flight, relatively long wings, dark cap, whitish supercilium, and wedge-shaped tail tip.

[e] Adult female. Similar to adult male but browner on back and more coarsely barred on underparts. Uppertail can show distinct dark barring (younger adults?).

[f] Juvenile *buteoides*. Far northern birds are overall paler.

[g] Adult. Note wide whitish supercilium, boldly barred underparts, unmarked secondaries, white undertail coverts, and wedge-shaped tip of tail.

[h] Juvenile. Long wings are tapered. Long tail shows irregular banding and wedge-shaped. Note two rows of buffy spots on upperwing coverts.

[i] Juvenile. Underparts are heavily streaked. Long, tapered wings are heavily barred below. Underparts usually fade to creamy by winter. Undertail coverts are streaked.

Similar species: Eurasian Sparrowhawk, Gyrfalcon

2. Eurasian Sparrowhawk *Accipiter nisus*   p. 131; photos, p. 320

Small slender hawk. Underparts barred in all plumages.

[a] Adult male. Cheeks are rufous. Underparts are barred rufous (solid rufous breast often). Flight feathers are heavily barred. Note square corners of tail tip.

[b] Adult female. Upperparts are brownish-grey. Dark tail bands are distinct.

[c] Adult male. Upperparts are blue-grey. Dark tail bands are indistinct.

[d] Juvenile male. Underparts are barred but with streaking on upper breast. Note narrow pale supercilium.

[e] Juvenile female. Underparts are barred but with rufous arrow-head markings. Juvenile back and upperwing coverts are brown, with pale feather edges in fresh plumage. Note narrow pale supercilium and narrow white terminal tail band.

[f] Adult male. Upperparts are blue-grey. Cheeks and underparts barring are rufous. Lacks pale supercilium.

[g] Adult male. Some (older?) birds have red eyes.

[h] Adult female. Upperparts are blue-grey, often with a brownish cast. Can appear brown-backed like juvenile females. Underparts lack rufous markings. Some show rufous wash on rear cheeks.

[i] Adult female. Underparts and flight feathers heavily barred. Undertail coverts are unmarked.

Similar species: Northern Goshawk, Levant Sparrowhawk, Kestrel, Merlin

Plate 23

Plate 24. PERCHED BUZZARDS

Adults have dark eyes; juveniles have pale ones. Wingtips reach tail tip on adults and fall just short of tip on juveniles, except for longer winged Rough-legged Buzzard.

1. Common Buzzard *Buteo b. buteo*   p. 144; photos, p. 324

[a] Whitish adult. White head shows no dark eye-line. No rufous on body.
[b] Normal adult. Dark underparts show a pale 'U' between breast and belly.
[c] Normal juvenile. Dark underparts show a pale 'U' between breast and belly.
[d] Whitish juvenile. White head shows no dark eye-line. No rufous on body.
[e] Dark-bibbed adult. Underparts are whitish with dark breast forming a bib.

Similar species: Other buzzards

2. Steppe Buzzard *Buteo b. vulpinus*   p. 149; photos, p. 326

Smaller than Common Buzzard.

[a] Grey-brown adult. Like normal Common Buzzard adult but overall paler.
[b] Darker juvenile. Like normal Common Buzzard juvenile but overall paler.
[c] Darker rufous adult. Underparts and tail are rufous.
[d] Paler rufous adult. Underparts and tail are rufous.
[e] Paler juvenile. Underparts are lightly streaked.

Similar species: Other buzzards

3. Long-Legged Buzzard *Buteo rufinus*   p. 154; photos, p. 328

Largest buzzard of the Western Palearctic, with relatively larger heads and beaks.

[a] Paler juvenile. Head is paler than body, with rufous eye-line. Legs and flanks, are rufous. Tail is banded only on lower half.
[b] Paler adult. Head is pale, and upperparts are rufous. Rufous tail is unbanded.
[c] Darker juvenile. Underparts are marked heavier on belly than breast. Tail is banded throughout.
[d] Darker adult. Underparts are marked heavier on belly than breast.

Similar species: Other buzzards

4. Rough-legged Buzzard *Buteo lagopus*   p. 160; photos, p. 364

Legs are feathered down to toes. Wingtips reach tail tip on juveniles; exceed on adults. Rough-legged Buzzards have relatively smaller beaks. Pale heads show dark eye-line.

[a] Adult male. Breast is more heavily marked than belly. Flanks and leg feathers are barred. Tail shows many dark bands. Note black face and orangish toes.
[b] Adult female. Flanks are uniform blackish-brown. Leg feathers are barred. Tail shows one dark band.
[c] Juvenile. Belly and flanks are uniform dark brown. Leg feathers are unmarked. Dusky tail tip lacks crisp dark band.

Similar species: Other buzzards

Plate 24

1[a] 1[b] 2[a] 2[b]
1[c] 2[c] 2[d] 2[e]
3[a] 1[d] 1[e] 4[b]
3[b] 3[c] 4[a]
3[d] 4[c]

Plate 25. PALER BUZZARDS FLYING

All usually show black carpal patches on underwings, except for most Steppe Buzzards, whitish Common Buzzards, and palest Long-legged Buzzards. Differences among species are mainly in markings and colour of body and tail and wing shape.

1. Rough-legged Buzzard *Buteo lagopus*   p. 160; photos, p. 364

Pale head shows dark eye-line. Longer winged than Common Buzzard.

[a] Juvenile. Upperparts are brown with buffy mottling. Note whitish primary panels and dusky band on tail tip.

[b] Juvenile. Appears pale below with wide, uniformly dark belly band, dark carpal patches, and dusky band on tips of flight feathers and tail.

Similar species: Other buzzards

2. Common Buzzard *Buteo b. buteo*   p. 144; photos, p. 324

Whitish variants. Whitish underparts are marked heavier on breast than on belly.

[a] Adult. Appears white below with dark carpal and flank patches.

[b] Adult. Shows white head and upperwing and uppertail coverts.

[c] Juvenile. Appears white below lacking dark carpal patches. Underparts are marked heavier on breast than on belly. Tail shows even-width narrow dark bands.

[d] Juvenile. Upperparts are brown with whitish mottling.

[e] Adult. Variant that shows dark bib but not dark carpal patches.

Similar species: Other buzzards

3. Steppe Buzzard *Buteo b. vulpinus*   p. 149; photos, p. 326

The Western Palearctic's smallest buzzard; paler individuals lack dark carpal patches.

[a] Pale juvenile. Pale brown upperparts lack whitish mottling. Tail shows even-width narrow dark bands.

[b] Pale rufous adult. Undersides are lightly marked with rufous. Rufous tail shows dark subterminal band (usually).

[c] Pale juvenile. Whitish underparts are lightly marked.

[d] Pale rufous adult. Upperwing coverts show rufous cast. Rufous tail shows dark subterminal band (usually).

Similar species: Other buzzards

4. Long-legged Buzzard *Buteo rufinus*   p. 154; photos, p. 328

The Western Palearctic's largest Buzzard. Relatively longer winged than others, more eagle-like.

[a] Pale adult. White head shows rufous eye-line. Flank and belly markings are heavier than those on breast. Pale rufous tail is usually unbanded.

[b] Pale juvenile. Upperparts are brown, with whitish primary panels. Tail shows only three or four narrow dark bands near tip. Note pale head.

[c] Pale adult. Back and upperwing coverts are rufous and contrast with darker flight feathers. Unbanded tail is pale rufous.

[d] Pale juvenile. Flank and belly markings are heavier than those of breast. Carpal patch is not complete. Tail shows dark bands only on lower half.

Similar species: Other buzzards

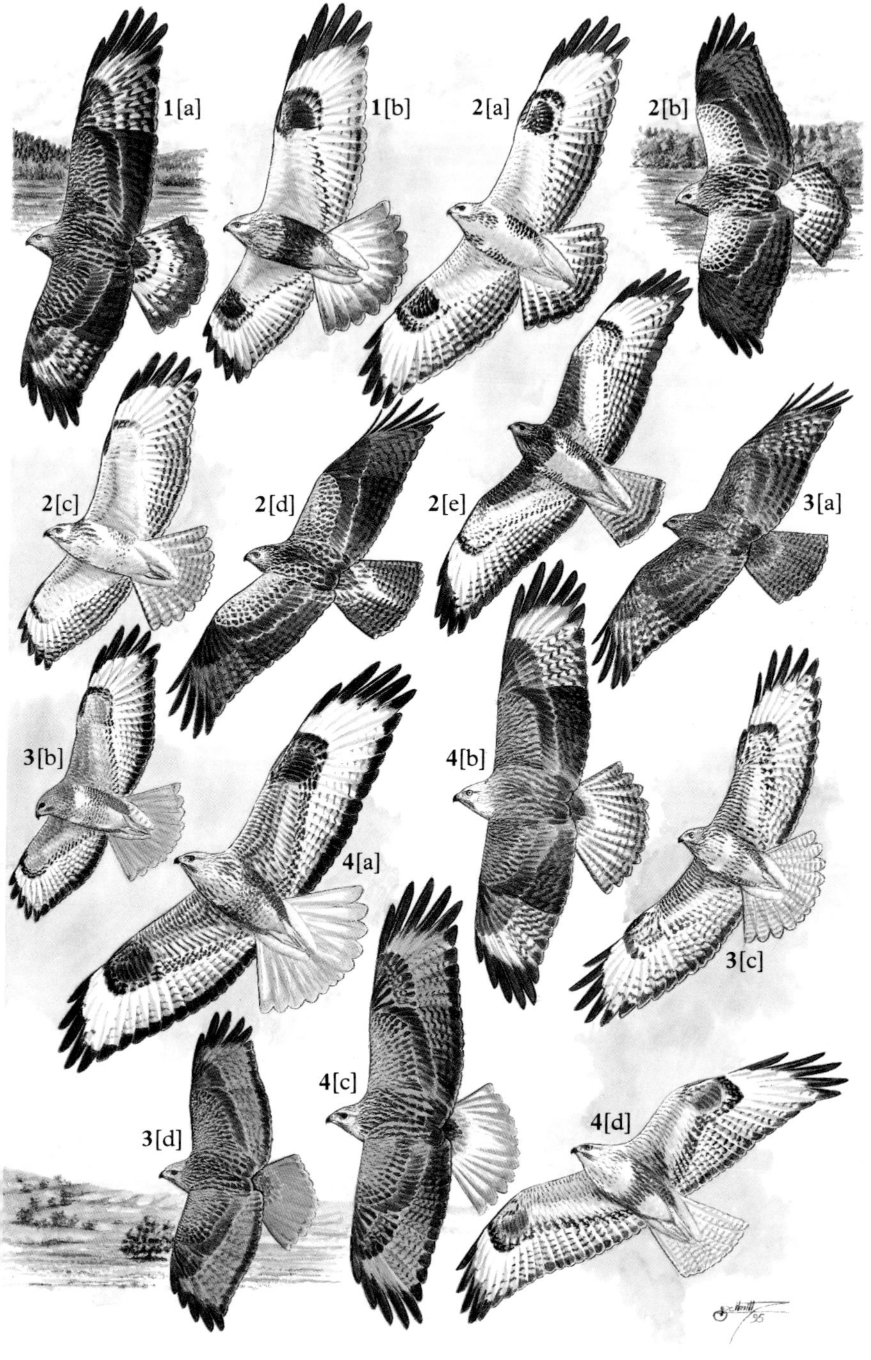
1[a]
1[b]
2[a]
2[b]
2[c]
2[d]
2[e]
3[a]
3[b]
4[b]
4[a]
3[c]
4[c]
3[d]
4[d]

Plate 26. DARKER BUZZARDS FLYING

All show black carpal patches on underwings, except for most Steppe Buzzards. Differences are mainly in markings and colour of body and tail and wing shape. Adults usually show a pale 'U' between breast and belly.

1. Rough-legged Buzzard *Buteo lagopus*   p. 160; photos, p. 364

Pale head shows dark eye-line. Longer winged than Common Buzzard.

[a] Adult male. Upperparts are heavily mottled. Tail has many dark bands.
[b] Adult male. Breast is more heavily marked than belly; flanks are barred.
[c] Adult female. Flanks are uniform blackish-brown. Tail has single dark band.
[d] Adult female. Upperparts are mottled. Tail shows one dark band.

Similar species: Other buzzards

2. Common Buzzard *Buteo b. buteo*   p. 144; photos, p. 324

Dark overall with pale 'U' between breast and belly.

[a] Adult. Shows numerous narrow dark tail bands, subterminal band wider, and wide dark band on trailing edge of wings.
[b] Adult. Dark above; uppertail shows numerous narrow darker bands.
[c] Juvenile. Shows numerous equal-width dark tail bands and narrow dusky band on trailing edge of wings.
[d] Juvenile. Dark above; uppertail shows numerous equal-width dark tail bands.

Similar species: Other buzzards

3. Steppe Buzzard *Buteo b. vulpinus*   p. 149; photos, p. 326

The Western Palearctic's smallest Buzzard; usually lacks dark carpal patches.

[a] Grey-brown adult. Like normal Common Buzzard adult but lacking blackish tones.
[b] Rufous adult. Upperside of rufous tail shows numerous dark bands.
[c] Rufous adult. Undersides of rufous with pale 'U' between breast and belly.
[d] Juvenile. Dark above; uppertail shows numerous equal-width dark tail bands.

Similar species: Other buzzards

4. Long-legged Buzzard *Buteo rufinus*   p. 154; photos, p. 328

The Western Palearctic's largest Buzzard. Relatively longer winged than others. Flank and belly markings heavier than those of breast.

[a] Juvenile. Flank and belly markings heavier than those of breast. Numerous equal-width dark tail bands.
[b] Adult. Flank and belly darker than breast. Rufous tail can show faint band.
[c] Juvenile. Upperparts lack rufous cast. Tail shows many equal-width dark bands.
[d] Adult. Back and upperwing coverts are rufous. Rufous tail lacks banding.

Similar species: Other buzzards

Plate 26

Plate 27. DARK- AND RUFOUS-MORPH BUZZARDS

Dark-morph buzzards have uniformly dark bodies and wing and tail coverts; similarly, rufous-morph birds are uniformly rufous there. Differences are mainly in markings and colour of tail and wing length and shape.

1. Common Buzzard *Buteo b. buteo* p. 144; photos, p. 324

Shown for comparison; does not have a true dark morph.

[a] Adult. Even darkest adults always show the pale 'U' on underparts.

[b] Adult. Similar to dark-morph Steppe Buzzard on upperparts.

2. Steppe Buzzard *Buteo b. vulpinus* p. 149; photos, p. 326

The Western Palearctic's smallest Buzzard. Relatively shorter winged than Long-legged Buzzard.

[a] Dark-morph adult. Body and coverts are uniformly black. Undersides of secondaries rather whitish.

[b] Dark-morph adult. Upperparts similar to those of dark-morph Long-legged Buzzard.

[c] Rufous-morph adult. Underparts are uniformly rufous, and rufous tail has narrow dark subterminal band.

[d] Rufous-morph adult. Back and upperwing coverts have a rufous-buffy cast.

[e] Dark-morph juvenile. Body and coverts are uniformly black. Pale tail shows numerous equal-width narrow dark bands.

[f] Dark-morph juvenile. Pale tail shows numerous equal-width narrow dark bands.

3. Long-legged Buzzard *Buteo rufinus* p. 154; photos, p. 328

The Western Palearctic's largest Buzzard. Relatively longer winged than others.

[a] Dark-morph adult. Body and coverts are uniformly black. Undersides of secondaries rather heavily marked. Whitish undertail shows wide dark subterminal band.

[b] Dark-morph adult. Upperparts similar to those of dark-morph Steppe Buzzard.

[c] Dark-morph adult. Alternate unbanded tail.

[d] Dark-morph juvenile. Body and coverts are uniformly dark brown. Undersides of secondaries are darkish. Dark tail shows four narrow white bands.

[e] Rufous-morph adult. Underparts are entirely rufous, darker on belly, with black carpal patches. Rufous tail shows many dark bands, subterminal widest.

[f] Rufous-morph adult. Back and upperwing coverts have a rufous-buffy cast. Rufous tail has many dark bands, subterminal widest.

[g] Dark-morph juvenile. Dark tail shows four narrow white bands.

4. Eurasian Honey Buzzard *Pernis apivorus* p. 20; photos, p. 290

Shown completely on Plates 6 and 7. One figure shown here for comparison.

[a] Dark-morph juvenile. Carpal patches appear darker than other coverts. Note pale primary patches and tail pattern with three darker bands.

Plate 27

Plate 28. LESSER SPOTTED EAGLE

Lesser Spotted Eagle *Aquila pomarina* p. 166; photos, p. 330

Small dark *Aquila* eagles whose wings and tails are relatively shorter than those of large *Aquila*. Spotted eagles have round nostrils. Underwing coverts usually appear paler than flight feathers. Sexes are alike in plumage. Wingtips almost reach tail tip on perched eagles.

[a]   Head-on silhouette. Wings are held level to wrist with wingtips drooped.

[b]   Juvenile. Body and underwing coverts are warm brown and appear paler than flight feathers. Juveniles show dark barring on flight feathers.

[c]   Distant adult. Classic silhouette. Note paler underwing coverts.

[d]   Adult. Body and wing coverts brown, with faintly barred flight feathers darker than underwing coverts. Note buffy tips on brown undertail coverts.

[e]   Second winter. Similar to juvenile, but new secondaries are longer and body and coverts appear somewhat mottled.

[f]   Distant adult. Upperwing coverts usually appear paler than back. Note white *Aquila* patches on upperwings.

[g]   Third winter. Similar to juvenile [h] and second winter [i], but new secondaries are longer and show less distinct barring. Note brown mottling on white undertail coverts.

[h]   Second winter. Similar to juvenile [b], but fresh longer secondaries noticeable. White spots on upperwing coverts hardly noticeable.

[i]   Juvenile. Juveniles from above show rufous nape patch, more uniform colour on upperparts, some small spots on upperwing coverts, and *Aquila* patches on upperwings.

[j]   Adult. Upperwing coverts are warm brown and appear paler than brown back. Note pale yellow eye and dark spotting in white uppertail coverts.

[k]   Juvenile. Appears overall warm brown, with rufous nape patch, small spots on upperwing coverts, and, usually, pale streaks on breast.

[l]   Adult. Adults have yellow eyes, upperwing coverts paler than backs, and buffy tips on brown undertail coverts. Note 'stove pipe' legs.

[m]   Juvenile. This individual shows heavy spotting on wing coverts, as well as spots on lower back and nape. Note rufous nape patch and 'stove pipe' legs.

[n]   Second winter. Similar to juvenile [m], but spotting on wing coverts is not as noticeable. Note fresh secondaries with wide white tips are longer than worn ones. Some show rufous nape patch.

Similar species: Greater Spotted Eagle, Steppe Eagle

Plate 28

[a]

[c]

[d]

[b]

[e]

[f]

[g]

[h]

[i]

[j]

[l]

[k]

[m]

[n]

Plate 29. GREATER SPOTTED EAGLE

Greater Spotted Eagle *Aquila clanga*   p. 171; photos, p. 332

Small dark *Aquila* eagles whose wings and tails are relatively shorter than those of large *Aquila*. Spotted eagles have round nostrils. Underwing coverts usually appear darker than flight feathers. Sexes are alike in plumage. Wingtips almost reach tail tip on perched eagles.

[a]   Head-on silhouette. Wings are held level to wrist with wingtips sharply drooped.

[b]   Adult. Overall dark brown, usually with somewhat paler grey-brown flight feathers, which contrast with darker underwing coverts. Occasionally underwings appear uniformly dark, and a few adults show slightly paler coverts. Some adults show white spotting on the undertail coverts.

[c]   Distant adult. Appears overall dark, often with a somewhat vulture-like S-curve on the trailing edge of wings.

[d]   Gliding. Note sharply bowed wings, with wrists up and wingtips down.

[e]   Third winter. Similar to juvenile and second winter but new secondaries are longer and darker, and usually lack dark barring. Dark underwing coverts contrast with paler flight feathers. Undertail coverts are mottled white and dark. This individual shows a thin white line through underwing (greater secondary coverts), a feature shown by some eagles of each age class.

[f]   Second winter. Similar to juvenile, but new feathers appear darker than old faded ones. New secondaries appear longer due to lack of wear. Dark underwing coverts contrast with paler flight feathers.

[g]   Juvenile. Underparts show a range or variation in amount and location of streaking, from uniformly streaked, as in this bird, to a pale breast patch in [e], to much paler belly in [n]. Pale secondaries contrast with darker underwing coverts and are narrowly barred dark, except for a wide area on tips.

[h]   Adult. Appears overall dark above, except for white *Aquila* patches and white in uppertail coverts. Upperwing coverts often appear a bit paler than flight feathers.

[i]   Second winter. Similar to juvenile, but new feathers are darker. New secondaries appear longer due to lack of wear.

[j]   Juvenile. This individual represents the heavily spotted extreme.

[k]   Juvenile. Juveniles show much white spotting on upperwing coverts and have all white uppertail coverts. Dark tails usually show a few pale bands near tip.

[l]   Adult. Overall dark brown, including undertail coverts. Note white on 'stove pipe' leg feathers.

[m]   Juvenile. This individual represents the lightly spotted extreme. Note the pale breast patch, ragged nape, and 'stove pipe' leg feathers.

[n]   Juvenile. This individual has very pale belly, leg feathers, and undertail coverts. Note white on 'stove pipe' leg feathers.

Similar species: Lesser Spotted Eagle, Steppe Eagle, Eastern Imperial Eagle

Plate 29

[a]

[b]

[c]

[d]

[e]

[f]

[g]

[h]

[i]

[j]

[k]

[l]

[m]

[n]

Plate 30. SPOTTED EAGLES

Similar plumages of Lesser and Greater Spotted Eagles are shown on this plate.

1. Greater Spotted Eagle *Aquila clanga*

Shown in detail on Plate 29

[a] Adult (darker). Overall blackish-brown. Some show a narrow white line through centre of underwing, recalling Steppe Eagle, but secondaries are unbarred.

[b] Adult (browner). Browner adults show underwing coverts paler than flight feathers, and appear much like Lesser Spotted Eagles. Note differences in wing shape, blackish carpals, markings on undertail coverts, and unbarred secondaries.

[c] *Fulvescens*-morph juvenile. Head, back, and upperwing coverts vary from rufous to straw-coloured, depending on amount of sun fading. Often appears similar to juvenile Eastern Imperial Eagle, but lower back is dark; only uppertail coverts are white.

[d] *Fulvescens*-morph juvenile. Head, body, and wing coverts vary from rufous to straw-coloured, depending on amount of sun fading. Often appears similar to juvenile Eastern Imperial Eagle, but underparts lack streaking and contrastingly paler inner primaries.

[e] Adult. Similar from above to adult Lesser Spotted Eagles, but back is usually same colour as upperwing coverts. *Aquila* patches on primaries show white shafts rather than white patches.

[f] *Fulvescens*-morph juvenile. Fresh plumage juveniles are rufous to rufous-buff. No adult specimens or records could be found.

[g] Older *fulvescens*. Darker rufous on head and underparts, with some black streaks on belly.

[h] Juvenile. Some juveniles have a rufous nape patch and appear similar to juvenile Lesser Spotted Eagles, but they are overall darker, with larger head and ragged crest, and usually have white on tarsi feathers (not shown here).

[i] Adult (paler). There is an overlap in colour with adult Lesser Spotted Eagles. Note larger head, more ragged crest, brown eye, and white patches on tarsi feathers.

2. Lesser Spotted Eagle *Aquila pomarina*

Shown in detail on Plate 28

[a] Adult. Similar in underwing pattern to some adult Greater Spotted Eagles, but note narrower wings, faint barring on flight feathers (if present), pale inner primaries, and buffy tips of brown undertail coverts.

[b] Adult. From above it is similar to adult Greater Spotted Eagles, but upperwing coverts are paler than the back. *Aquila* patches show white patches rather than white shafts.

[c] Adult (darker). There is an overlap in colour with adult Greater Spotted Eagles. Note smaller head with smooth nape, yellow eyes, all brown tarsi feathers, and pale tips on undertail coverts.

[d] Juvenile. Some juveniles are as heavily spotted as some juvenile Greater Spotted Eagles, but they are overall paler brown, with smaller head and smooth nape, and have all-brown tarsi feathers (not shown here).

3. Steppe Eagle *Aquila nipalensis*

Shown completely on Plates 31 and 32; figure here for comparison

[a] Fourth winter. First plumage adults are similar to adult Greater Spotted Eagles, but note heavily barred flight and tail feathers and relatively longer wings.

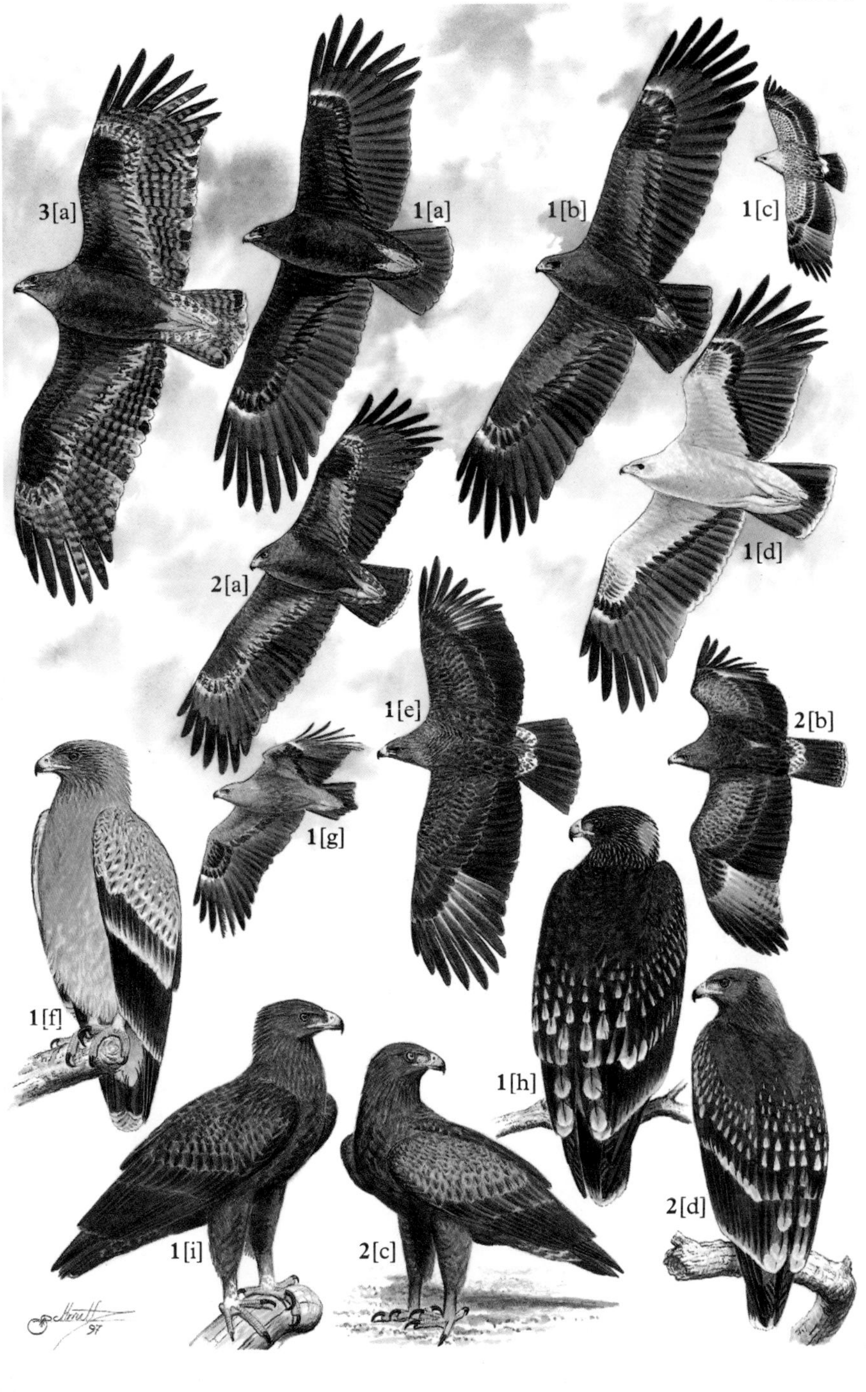
3[a]
1[a]
1[b]
1[c]
1[d]
2[a]
1[e]
2[b]
1[g]
1[f]
1[h]
2[d]
1[i]
2[c]

Plate 31. STEPPE EAGLE

Steppe Eagle *Aquila nipalensis*   p. 182; photos, p. 334

Steppe Eagles have noticeable wide gapes. Their tails are all dark, lacking any extensive white areas, usually showing narrow greyish banding. Eagles in the first three annual plumages usually show a wide pale band across each underwing. Eyes are dark, except for adults. Many individuals show a small white slash on lower back.

[a] Adult. Typical adult showing overall dark plumage, except for variable-sized rufous nape patch and buffy throat. Note horizontal attitude of Steppe Eagles when perched on the ground and that wingtips reach tail tip. Adult eye is yellow with brown flecking.

[b] Juvenile. Darker greyish-brown individual showing fresh even-width wide creamy tips of wing coverts and secondaries. Plumage of juveniles is uniform in colour, varying from pale buff to this dark.

[c] Second winter. One-year-old plus eagles are identical in plumage to juveniles but can be distinguished by wider creamy tips on the few new replacement coverts, longer new secondaries, and mottled appearance of body and coverts due to mix of old faded and new darker feathers. This individual is very pale.

[d] Third winter. Two-year-old plus eagles are darker than younger eagles but paler than adults, appear more mottled, and lack wide creamy tips of coverts, secondaries, and tails.

[e] Juvenile. Juveniles have uniformly coloured plumage on body and coverts. Most show a wide pale band across the middle of underwing. Most non-adults show pale inner primaries. Pale undertail coverts contrast with darker belly. Somewhat paler inner three primaries often form pale wedge on underwing.

[f] Second winter. One-year-old eagles appear much like juveniles except for some new longer secondaries and tail feathers and mottled body and coverts. Paler inner primaries are not always visible.

[g] Juvenile. Some immature eagles lack pale greater underwing coverts and show no pale band on underwings. But note even-width pale band on tips of secondaries and tail feathers of juveniles. A few individuals lack dark barring on secondaries.

[h] Juvenile. Upperside of juvenile shows narrow even-width bands on tips of greater coverts and tips of secondaries and tail feathers.

[i] Juvenile. Wing cutout showing underwing of variant having only primary coverts with wide creamy tips, forming a large pale comma at wrist.

[j] Third winter. Similar to juveniles and first summer eagles but lacking pale tips on upperwing coverts. New secondaries and tail feathers have at most narrow pale tips. Note pale patch on lower back. Somewhat paler inner three primaries often form pale wedge on upperwing. Appears similar to Lesser Spotted Eagles because of pale upperwing coverts.

[k] Head-on silhouette. Wings are drooped from the wrists outward.

Similar species: Lesser Spotted Eagle, Greater Spotted Eagle, Golden Eagle, Eastern Imperial Eagle

Plate 31

Plate 32. STEPPE AND TAWNY EAGLES

1. Steppe Eagle *Aquila nipalensis*   p. 182; photos, p. 334

Steppe Eagles have noticeable wide gapes. Their tails are all dark, lacking any extensive white areas, usually showing narrow greyish banding. Eagles in the first three annual plumages usually show a wide pale band across each underwing. Eyes are dark, except for adults. Many individuals show a small white slash on lower back.

[a] Third winter. Similar to juvenile and first summer birds but overall darker and more mottled. Some tips of greater underwing coverts have dark markings, but wide pale band through underwings is still noticeable. New secondaries and tail feathers do not have wide pale tips.

[b] Fourth winter. Darker overall than previous plumages, almost adult-like, with a less distinct narrower pale band visible on underwings. New secondaries and tail feathers have wide dark tips, and undertail coverts show dark barring.

[c] Second winter. Similar to juvenile but head, back, and upperwing coverts appear more mottled, less uniform. New secondaries, greater coverts, and tail feathers are longer than retained ones and show side pale tips.

[d] Fifth winter. First adult plumage often shows some retained immature characters in greater underwing coverts, secondaries, tail, and undertail coverts.

[e] Adult. Overall uniform dark brown, except for rufous nape patch and pale *Aquila* patches at base of primaries. Some show a small white streak on lower back. Greyish tail banding often noticeable.

[f] Fourth winter. Cutout of tail showing whitish uppertail coverts with dark markings.

[g] Adult. Overall dark brown except for buffy throat and boldly marked flight and tail feathers, which have wide dark terminal bands. Black carpal patches are usually noticeable.

Similar species: Lesser Spotted Eagle, Greater Spotted Eagle, Golden Eagle, Eastern Imperial Eagle

2. Tawny Eagle *Aquila rapax*   p. 177

Figures are also on Plates 33, 34, and 47.

The vagrant Tawny Eagle is usually rufous-tawny on head, body, and coverts, but some are grey-brown or creamy and overlap in colour with Steppe Eagles. Gape is noticeably smaller than that of Steppes; this is usually visible only on perched eagles. They never show the pale band across the underwings of non-adult Steppe Eagles.

[a] Juvenile. Similar to juvenile Steppe Eagle but pale primary patch is larger and more noticeable and secondaries and tail lack bold dark barring.

[b] Juvenile. Creamy individual. Secondaries and tails of first two plumages lack bold dark barring; secondaries are contrastingly darker than primaries and coverts. Always lacks wide pale bar across underwings, and undertail coverts are not contrastingly paler than belly.

Similar species: Steppe Eagle, Eastern Imperial Eagle

Plate 32

Plate 33. SPANISH IMPERIAL EAGLE

1. **Spanish Imperial Eagle** *Aquila adalberti* p. 188; photos, p. 336

Large *Aquila* that have long necks and heads and relatively long tails. They take 4 to 5 years to reach adult plumage. The juvenile and second winter plumages are different from those of adults. Sexes are alike in plumage. Wingtips almost reach tail tip on perched eagles.

[a] Head-on silhouettes. Wings are held level in soaring, more like a sea eagle than other *Aquila* eagles. But they sometimes glide with wings cupped.

[b] Adult. Back and upperwings are dark blackish-brown, except for white upper scapulars and marginal wing coverts that form white leading edges of inner wings. Head appears mostly straw-coloured, and tail shows much narrow greyish barring and dark tip.

[c] Juvenile. Colour of head, body, and coverts of older juveniles fades to creamy. Note pale inner primaries contrasting with dark secondaries.

[d] Adult. Dark blackish-brown overall except for white marginal wing coverts that form the white leading edges of inner wings. Note faint greyish barring on primaries. Edges of wings are rather straight. Straw-coloured cheeks set off dark throat.

[e] Second winter. Like juvenile but with mottled body and upperwing coverts and new secondaries longer and with greyish cast. Note pale inner primaries contrasting with dark secondaries.

[f] Juvenile. Has rufous-buff head, body, and wing coverts. Note pale inner primaries that contrast with darker secondaries and lack of body streaking. Nearly identical to juvenile Tawny Eagle, but with relatively longer wings.

[g] Third winter. Appears intermediate between juvenile and adult with a mix of new dark and old paler feathers. Note dark throat set of by straw-coloured cheeks.

[h] Juvenile. Juveniles in fresh plumage are dark rufous-buff overall except for dark flight and tail feathers.

[i] Second winter. Like juvenile but with short brownish streaking on body and mottling on upperwing coverts. Note broad white tips on longer new secondaries and greater upperwing coverts.

[j] Adult. Back and upperwing coverts are dark blackish-brown, except for white upper scapulars and lesser upperwing coverts; latter form white 'shoulders' on perched adult. Head appears mostly straw-coloured, except for dark forehead and throat. Tail shows much narrow greyish barring and dark tip.

Similar species: Golden Eagle, Tawny Eagle

2. **Tawny Eagle** *Aquila rapax*

Shown here for comparison. Other figures on Plates 32, 34, and 47.

[a] Juvenile. Nearly identical to juvenile Spanish Imperial Eagle in shape and pattern but is somewhat smaller, with relatively shorter wings.

Plate 33

1[a]
1[b]
1[c]
1[d]
1[e]
1[f]
1[g]
1[h]
1[i]
1[j]
2[a]

Plate 34. EASTERN IMPERIAL EAGLE

1. Eastern Imperial Eagle *Aquila heliaca*   p. 192; photos, p. 338

Large *Aquila* that have long necks and heads and relatively long tails. They take 4 to 5 years to reach adult plumage. The juvenile and second winter plumages are different from those of adults. Sexes are alike in plumage. Wingtips almost reach tail tip on perched eagles.

[a] Head-on silhouettes. Wings are held level in soaring, more like a sea eagle than other *Aquila* eagles. But they sometimes glide with wings cupped.

[b] Second winter. Like juvenile but with wide white tips on new longer secondaries and greater upperwing coverts. Note pale head, white extending up from uppertail coverts to lower back, and pale inner primaries contrasting with dark secondaries.

[c] Juvenile. Has creamy-buff head and heavily mottled, creamy-buff back and upperwing coverts. Note white extending up from uppertail converts to lower back and pale inner primaries that contrast with dark secondaries.

[d] Adult. Dark blackish-brown overall except for pale undertail coverts and (sometimes) greyish barring on flight feathers. Edges of wings are rather straight. Straw-coloured cheeks set off dark throat, and tail shows numerous narrow greyish bands and dark tip.

[e] Adult. Back and upperwings are dark blackish-brown, except for a few white scapulars. Head appears mostly straw-coloured, and tail shows numerous narrow greyish bands and dark tip.

[f] Juvenile. Juveniles have creamy-buff heads, body, and wing coverts, with dark streaking that begins on the neck and usually ends abruptly on belly. Dark secondaries contrast with paler inner primaries and wing coverts.

[g] Third winter. Appears intermediate between juvenile and adult with a mix of new dark and old paler feathers. Note dark throat set off by straw-coloured cheeks and pale undertail coverts.

[h] Juvenile. Juveniles are creamy-buff overall with dark brown streaking on underparts that begins on the neck and ends abruptly on belly in a straight line. Note the pale head.

[i] Third winter. Appears intermediate between juvenile and adult with a mix of new dark and old paler feathers. Note dark throat set off by straw-coloured cheeks and pale undertail coverts.

[j] Adult. Back and upperwings are dark blackish-brown, except for a few white scapulars. Head appears mostly straw-coloured, except for dark forehead and throat. Tail shows numerous narrow greyish bands and dark tip.

Similar species: Golden Eagle, Greater Spotted Eagle, Steppe Eagle

2. Tawny Eagle *Aquila rapax*

Shown here for comparison. Other figures on Plates 32, 33, and 47.

[a] Juvenile. Similar in faded plumage to juvenile Eastern Imperial Eagle in shape and pattern but lacks streaked underparts, is somewhat smaller, and has relatively shorter wings.

Plate 34

Plate 35. GOLDEN EAGLE

Golden Eagle *Aquila chrysaetos*   p. 197; photos, p. 340

Large dark *Aquila* that have golden napes and noticeably longer tails. They take 4 to 5 years to reach adult plumage. The juvenile plumage is distinct, whereas the older immature plumages are similar to each other. Sexes are alike in plumage. Wingtips almost reach tail tip on perched eagles.

[a] Head-on silhouette. Wings are held above the horizontal in a dihedral while gliding or soaring, different from any other *Aquila* eagle.

[b] Juvenile. Body and coverts appear uniform dark blackish-brown. White patches in wings are variable in size; some juveniles show none. Dark band on tail shows no grey markings. Tarsi feathers are usually white. Nape is same colour as that of older Eagles.

[c] Juvenile. Body and coverts appear uniform dark blackish-brown. Dark band on tail shows no grey markings. Nape of recently fledged eagle is more orangish. Juveniles lack tawny upperwing bars of older Goldens. White patches on upperwings differ from the *Aquila* patch of other eagles; many juveniles do not show these.

[d] Older immature. Similar to juvenile but with mottled body and wing coverts. New secondaries are shorter, show irregular greyish banding, and have dark tips. New tail feathers show a more irregular border between white base and dark tip. Tarsi feathers are now tawny.

[e] Older immature. Similar to juvenile but with mottled body and wing coverts. New secondaries are shorter and have dark tips. New tail feathers have greyish markings in dark band on tip and more irregular border between white base and dark tip. Note buffy upperwing bars.

[f] Adult. Dark brown below, sometimes with rufous or whitish streaks on breast. Underwings and undertail show irregular greyish barring when seen in good light, with dark band noticeable on trailing edges of underwings and undertail.

[g] Adult. Dark above, with golden nape and buffy upperwing bars usually visible. Irregular greyish barring on uppersides of flight feathers and uppertail are seen in good light. Note the lack of the *Aquila* patches on upperwings.

[h] Second winter. Some eagles replace only the central tail feathers and inner three primaries, but no secondaries, in their first annual moult; as a result, they appear much like juveniles, but note the mottled body and wing coverts, due to a mix of new and old feathers, and buffy band across upperwings.

[i] Fifth winter. First plumage adults can show whitish rays on the sides of some flight and tail feathers.

[j] Adult. Overall dark brown with golden nape and buffy wing bar. Eye is usually yellow-brown, but is variable. Some adults show rufous breast patches.

[k] Juvenile. Compared to older eagles, juveniles are uniformly darker brown, lacking the buffy wing bars. Eyes are dark brown. Tail shows white base and dark tip, with a rather even boundary between them. Tarsi feathers are usually white.

Similar species: Eastern Imperial Eagle, Spanish Imperial Eagle, Steppe Eagle

Plate 35

Plate 36. BONELLI'S EAGLE

Bonelli's Eagle *Hieraaetus fasciatus* p. 206; photos, p. 342

Medium-sized pale eagles. Adults have white body and undertails, latter with wide dark terminal band. Sexes are alike in plumage. Wingtips do not reach tail tip on perched eagles.

[a] Adult. Typical adult showing dark brown uppersides with a variably sized white triangle on back and wide dark subterminal band on grey tail.

[b] Juvenile. Typical juvenile showing brown uppersides and faint narrow dark banding on paler brown tail. Primaries are paler than secondaries.

[c] Adult. Paler adult with sparse fine dark streaking on underparts and whitish mottling on undersides of flight feathers. Note wide black band through underwings and wide dark subterminal tail band.

[d] Second winter. Similar to juvenile but with heavy dark streaking on darker rufous undersides and many new flight and tail feathers that have wide dark tips.

[e] Adult. Darker adult with more heavily streaked underparts and darker undersides of flight feathers. Note wide black band through underwings and wide dark subterminal tail band.

[f] Third winter. Typical first plumage adult with barring on undersides of paler flight feathers and wider dark streaks on underparts. Note wide black band through underwings and wide dark subterminal tail band.

[g] Juvenile. Fresh plumage juveniles are bright rufous-brown below, usually with a narrow black line through centre of underwings. Note that darker secondaries form somewhat noticeable patches on underwings.

[h] Juvenile. Typical juvenile perched. Note brown tail with narrow dark banding.

[i] Second winter. Similar to juvenile but with wide dark streaks on darker rufous underparts and greyish tail with dark subterminal terminal band.

[j] Adult. Typical adult. Some lack white triangle on back. Note the heavily mottled leg feathers and grey tail with wide dark subterminal band.

[k] Juvenile. Juveniles in late winter and spring usually show faded to creamy or even whitish underparts and underwing coverts. Some juveniles lack the black line through underwings. Note that darker secondaries form somewhat noticeable patches on underwings.

Similar species: Booted Eagle, Short-toed Snake Eagle

Plate 36

[a]

[b]

[c]

[d]

[e]

[f]

[g]

[h]

[i]

[j]

[k]

Plate 37. BOOTED EAGLE

1. **Booted Eagle** *Hieraaetus pennatus* p. 210; photos, p. 344

Small polymorphic eagles. Adults and juveniles appear alike in the field. Sexes are alike in plumage. Wingtips almost reach tail tip on perched eagles.

[a] Head-on silhouette. Typical wing attitude with wrists up and tips down. White 'landing lights' are not always visible.

[b] Rufous morph. Underparts and lesser underwing coverts are rufous-brown. Note wide dark band through centre of underwings and pale inner primaries.

[c] Dark morph. Overall dark brown on underparts and underwing coverts. Inner primaries are usually paler. Darker individual. Note dusky band on tail tip.

[d] Light morph. White underparts and underwing coverts that contrast with dark flight feathers are distinctive. Note pale inner primaries. Underparts can show a variable amount of narrow dark streaking.

[e] Dark morph. Paler variation of dark morph in typical fast glide. Note rufous undertail coverts and dusky band on tail tip.

[f] All morphs. Uppersides of all morphs are similar, with pale bar across upperwings, pale scapulars, white greater uppertail coverts forming a pale 'U' above base of tail, and white 'landing lights'. Usually show dark face and pale crown and nape.

[g] Typical stoop. Booted Eagles usually catch prey at end of near vertical stoops.

[h] Juvenile. Recently fledged juveniles have dark brown eyes, which change to more amber coloration by their first spring. Median upperwing coverts have not faded.

[i] Dark morph. Underparts are dark brown. Adults have golden-yellow eyes.

[j] Rufous morph. Underparts are rufous-brown. Usually show dark face and pale crown and nape.

[k] Faded rufous morph. Rufous underparts often fade to creamy-buff.

[l] Light morph. Underparts are white.

Similar species: Bonelli's Eagle, Black Kite, Marsh Harrier, Egyptian Vulture

2. **Bonelli's Eagle** *Hieraaetus fasciatus*

Shown completely on Plate 36. Fresh plumage juvenile shown here for comparison with rufous-morph Booted Eagle.

[a] Juvenile. Appears similar to rufous-morph Booteds but is larger, has a narrower black line through underwings and relative broader wings, and lacks dusky band on tail tip.

3. **Black Kite** *Milvus migrans*

Shown completely on Plates 8 and 9. Darker kite depicted here for comparison with dark-morph Booted Eagle.

[a] Juvenile. Appears similar to dark-morph Booteds but is larger, shows square tail tip lacking dusky band, and has pale panel at base of outer primaries.

Plate 37

Plate 38. KESTREL MALES

1. **Common Kestrel** *Falco tinnunculus*   p. 221; photos, p. 348

Distinguished by single heavy moustache mark. Talons are dark.

[a] Adult male. Rufous back and upperwing coverts show dark diamond-shaped markings. Wingtips can reach dark subterminal band of central tail feathers.

[b] Adult male. Note two-toned pattern on upperwings and *rufous back with dark diamond-shaped markings*.

[c] Adult male. Note uniformly marked underwings.

Similar species: Lesser Kestrel, Red-footed Falcon, Eurasian Sparrowhawk

2. **Lesser Kestrel** *Falco naumanni*   p. 216; photos, p. 346

Wingtips reach dark subterminal band of central tail feathers. Talons are pale.

[a] Adult male. Head entirely is grey with buffy throat but no moustache mark, streaking on nape, or white cheeks. Note *grey greater secondary coverts. Bright rufous back is unmarked.* Rufous-buff underparts are unmarked or lightly spotted.

[b] First summer male. Appears much like adult male Common Kestrel, but note *unmarked back*, lack of dark eye-line, *pale talons*, and fainter moustache. Wingtips reach well on to dark subterminal band of central tail feathers.

[c] Adult male. Note *grey greater secondary coverts*. Projecting central tail feathers are suggestive but some Lessers lack them and a few Common Kestrels have them.

[d] First summer male. Similar to adult male Common Kestrel but back is unmarked and tail shows retained juvenile outer feathers.

[e] First summer male. Similar to adult male Common Kestrel but note that secondaries are less boldly marked than are underwing coverts. They usually show retained juvenile outer tail feathers.

[f] Adult male. Note *whitish, almost unmarked underwings* and pale to dark buffy-rufous body that is usually unmarked.

Similar species: Common Kestrel, Red-footed Falcon, Levant Sparrowhawk

3. **American Kestrel** *Falco sparverius*   p. 226

Vagrant from North America. They are more compact than European Kestrels. Note more distinctive face pattern, blue-grey upperwing coverts, rufous tail, and much less distinct two-toned pattern on upperwings. Proportions are more similar to those of male Merlins than of other Kestrels. All show two black spots, 'false eyes', on nape.

[a] Juvenile male. Juvenile male has whitish breast with fine dark streaks and completely barred upper back. Note one of the 'false eyes'.

[b] Adult male. Has unbarred upper back. Note blue-grey coverts and rufous tail.

[c] Juvenile male. Note undertail pattern and colour and row of white dots on trailing edges of underwings. (Underwing is shown in shadow to emphasize the row of white dots; thus it appears somewhat darker than normal.)

Plate 38

1[a]

3[a]

2[a]

2[b]

2[c]

2[d]

1[b]

3[b]

2[e]

1[c]

2[f]

3[c]

Plate 39. KESTREL FEMALES AND JUVENILES

1. **Common Kestrel** *Falco tinnunculus*   p. 221; photos, p. 348

Distinguished by narrow dark line behind eye and dark talons. Two-toned upperwing pattern is shared with Lesser Kestrel. *Wingtips do not reach dark subterminal band of central tail feathers* on perched falcons.

[a] Juvenile female. Juveniles have even-width dark brown bands across back and wing coverts. Note dark line behind eye. Juvenile females have wide dark tail bands.

[b] Adult female. Adult females have dark triangular marks on back and wing coverts. Many females show grey uppertail coverts with dark shaft streaks and a variable amount of greyish wash on uppertail. Dark tail bands are narrow.

[c] Juvenile female. Juveniles of both sexes are almost alike and similar to adult female but show wide even-width dark barring on backs and upperwing coverts. Males often have greyish cast to uppertail coverts and uppertail similar to that of adult females. Adult females and juveniles show tawny markings on primary coverts.

[d] Adult female. Underwings of Common Kestrels show little contrast between dark markings on coverts and undersides of secondaries.

[e] Adult female or juvenile head. Head is relatively larger, with longer, darker moustache mark and narrow *dark line behind eye*.

Similar species: Lesser Kestrel, Red-footed Falcon, Eurasian Sparrowhawk

2. **Lesser Kestrel** *Falco naumanni*   p. 216; photos, p. 346

Distinguished by *pale talons and lack of dark line behind eye*. Both kestrels can show wedge-shaped tail tips and two-toned upperwings. Adult female and juveniles are not easy to distinguish in the field. *Wingtips reach dark subterminal band of central tail feathers* on perched falcons.

[a] Juvenile. Juveniles have wide even-width dark barring on upperparts.

[b] Adult female. Adult females have dark triangular markings on back and upperwing coverts and blue-grey uppertail coverts.

[c] Adult female. Primary coverts of adult females and juveniles are completely dark brown. Compare with those of Common Kestrels (1c).

[d] Adult female. Similar to juvenile and adult female Kestrels but note less boldly marked secondaries that contrast with more darkly marked coverts, projecting central tail feathers, and fainter moustache.

[e] Adult female or juvenile head. Head is relatively smaller, shows fainter moustache mark, and *lacks dark line behind eye*. However, some may show a faint line.

Similar species: Common Kestrel, Red-footed Falcon

3. **American Kestrel** *Falco sparverius*   p. 226

Vagrant from North America. Note more distinctive face pattern and much less distinct two-toned pattern on upperwings. Proportions are more similar to male Merlins that to above kestrels; they are more compact than European kestrels. Adult and juvenile females are alike. All show black spots, 'false eyes', on nape.

[a] Female. Note more strongly marked head and blurry rufous breast streaks.

[b] Female. Note less distinct two-toned pattern on upper wings.

[c] Female. Note strong head pattern.

Plate 39

1[a]

3[a]

2[a]

2[b]

2[e]

1[b]

1[e]

1[c]

3[b]

2[c]

2[d]

1[d]

3[c]

Plate 40. RED–FOOTED AND AMUR FALCONS

1. Red-footed Falcon *Falco vespertinus* p. 229; photos, p. 350

On perched falcons, wingtips extend beyond tail tip. Juvenile in flight from above is shown on Plate 42 for comparison with juvenile Eurasian Hobby.

[a] Adult male. Appears overall dark grey except for bright rufous leg feathers and undertail coverts. Underwings appear rather uniform in colour, but glossy flight feathers can appear paler than matte coverts in some light.

[b] Adult male. *Cere, base of beak, eye-ring, and legs are red-orange. Note silvery primaries.*

[c] Adult female. Head and underparts are orange-buffy. Blue-grey upperparts are barred dark. Note small black eye patch and short black moustache.

[d] Juvenile (sexes alike). Head shows brown crown, small black eye patch, short black moustache, and whitish nape collar. Sandy-brown upperparts are barred dark brown, and creamy underparts have rufous-brown streaks. Tail shows even-width grey and black bands.

[e] First summer male. Juveniles returning in spring show some new adult body feathers and retained juvenile flight feathers and underwing coverts. Central tail feathers are usually replaced.

[f] Second winter male. Note retained juvenile flight feathers.

[g] Adult female. Head, underparts, and underwing coverts are orangish-buff. Note wide dark band on trailing edge of wings.

[h] First summer female. Similar to adult female but paler, with narrow black shaft streaking on underparts and retained juvenile wing coverts and tail.

[i] Adult female. Upperparts are blue-grey, with obvious orange-buff head.

[j] Juvenile. Underparts are heavily streaked. Note dark terminal band on wings.

[k] Juvenile. Kestrel-like: sandy back and upperwing coverts contrast with darker flight feathers. Tail shows even-width rufous-buff and dark brown bands.

[l] Adult male. Appears dark grey above, except for silvery flight feathers.

Similar species: Eurasian Hobby, Common Kestrel, Merlin, Eleonora's Falcon, Sooty Falcon

2. Amur Falcon *Falco amurensis* p. 234

Vagrant to the Western Palearctic from eastern Asia.

[a] Adult male. Like adult male Red-foot but with *white underwing coverts.* Perched (not shown) show pale grey cheeks with noticeable darker moustache stripes.

[b] Adult female. Similar to juvenile Red-foot but *streaking on underparts is blackish and flanks are barred.* Note wide dark subterminal tail band. Upperparts (not shown) are blue-grey.

Juvenile male (not illustrated). In spring (subadult) will have greyish body and dark streaking on breast.

Juvenile female (not illustrated). Similar in spring to adult female, as some moult on body takes place over winter. New central tail feathers may show wide subterminal band, but retained juvenile tail feathers will not.

Similar species: Red-footed Falcon

Plate 40

2[a]

1[a]

1[c]

1[d]

1[b]

1[f]

1[g]

1[e]

1[i]

1[h]

1[j]

2[b]

1[k]

1[l]

Plate 41. ELEONORA'S FALCON

1. Eleonora's Falcon *Falco eleonorae* p. 238; photos, p. 352

In flight, their long, narrow wings and long tails are distinctive. On perched falcons, wingtips extend a bit beyond tail tips. Occurs in two colour morphs: light is the most common (~75 per cent), and dark is fairly common (~25 per cent). Dark morph is further divided into darkish (~23 per cent) and all-dark (~2 per cent). See text for differences.

[a] Light-morph adult male. Head shows narrow dark moustache mark and small round white cheek patches. Underparts have sooty-rufous wash. Adult male has yellow cere and eye-ring. Note short tarsi.

[b] All-dark adult female. Overall blackish grey. Adult female has blue cere and yellowish eye-rings. Note short tarsi.

[c] Darkish adult. Appears like dark-morph adult but with dusky throat and cheeks and dark rufous wash or streaks on underparts. Paler individual.

[d] Light-morph Juvenile. Head shows narrow dark moustache and small round buffy cheek. Juveniles have brown backs and upperwing coverts with buffy edges.

[e] Light-morph adult male. Underwings are two-toned: darker coverts contrast with paler flight feathers. Underparts have sooty-rufous wash. Adult male has yellow cere and eye-ring. Note long tail with wedge-shaped tip.

[f] Light-morph juvenile. Underwings are somewhat two-toned. Creamy underparts are streaked on to belly. Light-morph juveniles have no markings on undertail coverts. Long tail has wedge-shaped tip.

[g] Darkish juvenile. Similar to but darker than light-morph juveniles, with more heavily marked underparts and barred undertail coverts.

[h] Darkish adult female. Overall dark. Underwings are two-toned: paler flight feathers contrast with darker coverts. Darker individual with rufous wash.

[i] Light-morph adult. Appears slate grey above. Long grey tail has wedge-shaped tip.

[j] Light-morph juvenile. Dark brown upperparts have buffy feather edges, and long tail has wedge-shaped tip. Note rounded buffy cheeks and dark nape.

[k] Darkish adult. Individual with white throat.

[l] All-dark juvenile. Like dark-morph adult, but with rufous feather edges. Cere and eye-ring are bluish. Undertail is dark with narrow pale bands.

Similar species: Eurasian Hobby and juveniles of Lanner Falcon, Peregrine, Red-footed Falcon, Sooty Falcon

2. Lanner Falcon *Falco biarmicus* p. 258; photos, p. 360

Lanner Falcon is shown in detail on Plate 44. Juvenile is shown here for comparison.

[a] Juvenile male. Streaking on underparts ends abruptly on belly, crown is pale, and dark eye-line is noticeable. Underwings are a bit more two-toned than are those of juvenile Eleonora's Falcons.

3. Peregrine *Falco peregrinus* p. 277; photos, p. 366

Peregrine is shown in detail on Plate 46. Juvenile is shown here for comparison.

[a] Juvenile female of race *brookei*. Compared with juvenile Eleonora's, Peregrines are heavier bodied and wider winged. Underwings are more uniform in colour.

Plate 41

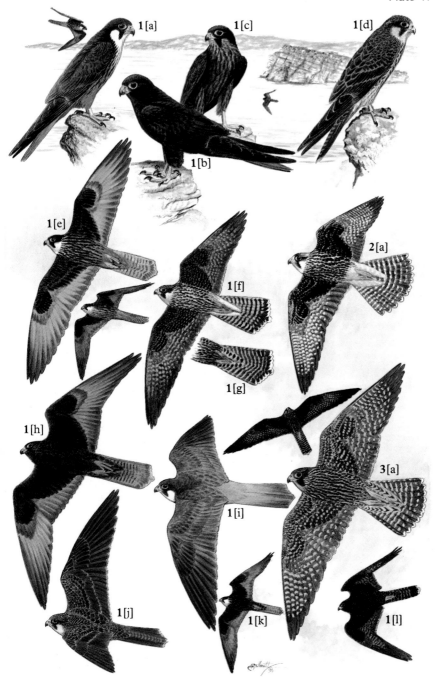

1 [a]

1 [c]

1 [d]

1 [b]

1 [e]

1 [f]

2 [a]

1 [g]

1 [h]

1 [i]

3 [a]

1 [j]

1 [k]

1 [l]

Plate 42. EURASIAN HOBBY AND SOOTY FALCON

1. Eurasian Hobby *Falco subbuteo*   p. 253; photos, p. 356

Medium-sized falcon. Wings are long and narrow; tail appears short. Always shows strong face pattern with two black moustache marks and pale sides of nape.

[a] Adult. Underparts are heavily streaked; underwings are uniformly dark (but can appear somewhat two-toned). Adults have rufous leg feathers and undertail coverts.

[b] Juvenile undertail cutout. Pale bands are more distinct than adult's.

[c] Distant adult. Classic glide profile.

[d] Adult. Upperparts of adults are dark blue-grey; their leg feathers and undertail coverts are rufous. Wingtips extend a bit beyond tail tip.

[e] Juvenile. Juveniles have creamy underparts with wide dark brown streaks; some extend from throat to lower belly. Wingtips extend just beyond tail tip.

[f] Juvenile. Uppersides are dark brown, with faint tail banding. Compare to juvenile Sooty Falcon (2b) and juvenile Red-footed Falcon (3a).

[g] Silhouette. Glides often with swift-like attitude, with wings depressed.

Similar species: Sooty Falcon, Red-footed Falcon, Eleonora's Falcon, Merlin, Peregrine

2. Sooty Falcon *Falco concolor*   p. 243; photos, p. 354

Medium-sized falcon. Wings are long and narrow. On perched falcons, wingtips extend beyond tail tip, which show projecting central feathers.

[a] Adult. Overall sooty grey. Tip of black tail is rather wedge-shaped.

[b] Juvenile. Uppersides are dark grey-brown. Tail shows little or no banding. Compare to juvenile Eurasian Hobby (1f) and juvenile Red-footed Falcon (3a).

[c] Adult female. Overall sooty grey. Tips of primaries and tail are black. Cere, face-skin, and legs of females are yellow.

[d] Juvenile. Orange-buff undersides are heavily streaked. Note two Hobby-like dark moustache marks. Wingtips extend beyond tail tip.

[e] Adult male. Overall sooty grey. Cere, face-skin, and legs of males are orangish. Wingtips extend well beyond tail tip.

[f] Juvenile. Orange-buff undersides are heavily streaked, with dusky band across breast. Flight feathers are darker than underwing coverts. Tail tip shows wide unbanded area and projecting central feathers.

Similar species: Eurasian Hobby, Red-footed Falcon, Eleonora's Falcon

3. Red-footed Falcon *Falco vespertinus*   p. 229; photos, p. 350

Red-footed Falcon is illustrated in detail on Plate 40. Juvenile is shown here for comparison.

[a] Juvenile. Sandy back and upperwing coverts contrast with dark brown flight feathers. Tail banding is distinct. Compare with 1[f] and 2[b].

Plate 42

1[b]
1[c]
1[d]
1[e]
1[a]
1[g]
3[a]
1[f]
2[b]
2[a]
2[c]
2[d]
2[e]
2[f]

Plate 43. MERLIN

Merlin *Falco columbarius*   p. 248; photos, p. 358

Small compact falcon. They show at most a *faint moustache mark*. On perched falcons, wingtips fall short of tail tips.

Three races are: (1) *F. c. aesalon* (widespread in northern Europe); (2) *F. c. subaesalon* (Iceland); and (3) *F. c. pallidus* (Central Asia).

1. *F. c. aesalon* usually shows faint moustache mark.

[a] Adult female. Upperparts are dark brown. Note faint whitish collar.

[b] Adult male. Upperparts are blue-grey. Note rufous collar and orangish legs.

[c] Adult female. Underwing appear dark. Tail shows wide buffy bands.

[d] Adult male. Upperparts are blue-grey, with wide black band on tail tip.

[e] Adult male alternate tail cutout. Shows multiple dark tail bands.

[f] Adult male. Underparts are rufous.

[g] Adult female. Upperparts are dark brown. Dark brown uppertail shows buffy bands.

[h] Juvenile male. Has hint of greyish cast to rump and uppertail coverts, which is lacking on juvenile female.

2. *F. c. subaesalon*. overall averages darker than *aesalon*, but with much overlap. Adult males are not separable in the field from *aesalon*.

[a] Adult female. Averages darker on back than adult female *aesalon* with narrower pale tail bands and lacking rufous feather edges on upper back.

[b] Adult female. Averages darker on back than adult female *aesalon* with narrower pale tail bands.

3. *F. c. pallidus* is overall much paler than *aesalon*. Moustache marks are faint or absent.

[a] Adult female. Upperparts are sandy brown with rufous cast. Note wide buffy tail bands.

[b] Adult male. Upperparts are pale blue-grey.

[c] Adult male. Upperparts are pale blue-grey.

[d] Adult female. Appears somewhat kestrel-like, but lacks two-toned upperwings. Note wide buffy tail bands.

[e] Juvenile male. Appears somewhat kestrel-like. Slight blue-greyish cast to rump and uppertail coverts (lacking in juvenile females).

Similar species: Eurasian Hobby, Red-footed Falcon, Common and Lesser Kestrels

Plate 43

Plate 44. LANNER FALCON

Lanner Falcon *Falco biarmicus* p. 258; photos, p. 360

Large slender falcon, with long wings and tail. Head shows long narrow dark moustache marks, dark eye-line, and dark forecrown. Wingtips reach or almost reach tail tip on perched falcons. Three races, two of which are identical, that occur in the Western Palearctic are: (1) *F. b. feldeggii*, European race that differs somewhat from North African races; (2) *F. b. tanypterus* and *F. b. erlangeri*, which are identical.

1. *F. b. feldeggii* Darkest race. Wingtips fall cms short of tail tip.

[a] Second winter. Like adult but underparts are more heavily marked, and rufous on crown and nape shows fine dark streaks.

[b] Adult male. Adults have ashy-grey upperparts. Adult *feldeggii* have barred leg feathers; males have rufous crowns and black foreheads.

[c] Juvenile. Underparts are heavily streaked. This juvenile shows dark crown.

[d] Juvenile. Upperparts are dark brown. Juveniles of all races appear similar.

[e] Juvenile. Underwings are two-toned: dark coverts contrast with paler flight feathers. *Note abrupt end of dark streaking on belly.*

[f] Adult. Pale underwings and dark barring on flanks are distinctive.

[g] Adult female. Adult female *feldeggii* have completely or near completely dark crowns, with rufous restricted to nape.

[h] Darker adult. Heavily marked greater coverts form dark band on underwings.

2. *F. b. tanypterus* and *F. b. erlangeri*. Similar to *F. b. feldeggii* but average paler and less heavily marked. Wingtips reach tail tip on perched adults and almost reach on perched juveniles.

[a] Adult Lanner head (lower) comparison with adult Barbary Falcon head (upper). Lanners show more distinct dark eye-lines, rufous on crown, and white cheeks.

[b] Second winter. Like adult but more heavily marked on underparts.

[c] Adult. Pale extreme showing creamy crown and lacking flank barring.

[d] Adult. Typical adult with rufous-buffy crown.

[e] Juvenile. Pale, lightly marked individual with completely white crown.

[f] Adult. Flanks are lightly marked, and underwings are pale.

[g] Cutout of adult head. Shows pale crown.

[h] Cutout of adult tail. Pale extreme.

Similar species: Peregrine, Barbary Falcon, Eleonora's Falcon, Saker Falcon

Plate 44

Plate 45. SAKER FALCON AND GYRFALCON

1. **Saker Falcon** *Falco cherrug*   p. 264; photos, p. 362

Large falcon with pale head, uniformly coloured crown, faint narrow moustache marks, and brown back. On perched falcons, wingtips fall short of tips of long tail. European Sakers are rather uniform, with little variation. However, escaped Asiatic Sakers, common in captivity in the Middle East, vary from dark to rufous to whitish.

[a] Juvenile. Juveniles' underparts are heavily streaked. Ceres and legs are blue-grey. Note pale head and whitish spots on tail.

[b] Juvenile. Underwings are two-toned: dark coverts contrast with paler flight feathers. Streaking on underparts extends on to lower belly.

[c] Adult. Adults' upperparts are brown. Underwings are uniformly pale. Note pale head and long wings and tail. Paler adult with sparse spotting on belly and flanks and fine streaking on breast.

[d] Adult. Darker adults are marked with rows of dark spots, appearing as streak on flanks and belly; their ceres, eye-rings, and legs are yellow. Note whitish spots on tail. Wingtips fall short of tail tip.

[e] Darker juvenile. Note darker head and unmarked tail.

[f] Cutout of tail. Lightly marked tail.

[g] Cutout of tail. Heavily spotted tail.

[h] Adult. Upperwings of Sakers are somewhat two-toned: dark brown outer half of wings contrasts with paler inner half. Spotting on tail forms pale bands.

Similar species: Lanner Falcon, Peregrine

2. **Gyrfalcon** *Falco rusticolus*   p. 268; photos, p. 364

The Western Palearctic's largest falcon, with relatively small head and robust breast. Wing and tail have broad bases. On perched falcons, wingtips fall short of tail tip.

[a] Juvenile. Underparts are heavily streaked with dark grey-brown. Underwings are two-toned: dark coverts contrast with paler flight feathers. Note broad bases of wings and tail. Cere, eye-rings, and legs are blue-grey.

[b] Adult. Slate-grey upperparts are heavily cross-barred pale grey. Note broad-based tail that can appear tapered. Cere, eye-rings, and legs are yellow.

[c] White-morph juvenile. Upperparts are brownish and whitish head shows no moustache marks. Occurs as vagrant from Greenland.

[d] Adult. Underparts consist of streaked breast, barred flanks, and spotted belly. Perched Gyrs appear broad-chested and small-headed. Note dark moustache mark on darkish head. Adults have yellow legs.

[e] Juvenile. Upperparts are dark greyish-brown; underparts are heavily streaked. Note dark moustache mark on darkish head. Perched Gyrs appear broad-chested and small-headed. Cere, eye-rings, and legs are blue-grey.

Similar species: Northern Goshawk, Peregrine

Plate 45

1[a]
1[b]
1[e]
1[f]
1[d]
1[c]
1[g]
1[h]
2[e]
2[b]
2[a]
2[c]
2[d]

Plate 46. PEREGRINE AND BARBARY FALCON

Peregrine *Falco peregrinus*   p. 277; photos, p. 366

Large falcon, with dark hood and moustache mark. Underwings are usually uniformly dark. On perched falcons, wingtips reach or almost reach tail tip. The Western Palearctic races are: (1) *F. p. peregrinus*, European race; (2) *F. p. calidus*, tundra race; (3) *F. p. brookei*, Mediterranean race. There is much intergrading in plumage characters among the three races.

1. *F. p. peregrinus*
[a] Adult male. Blackish head appears 'hooded' and shows wide moustache mark. Underparts have a rufous-buff wash, lighter on males.
[b] Juvenile female. Underparts are heavily streaked. Underwings appear dark.
[c] Adult female. Head appears hooded with wide moustache mark. Underparts show rufous-buff wash, heavier on females.

2. *F. p. calidus*. Paler overall than nominate, with narrow moustache marks.
[a] Adult female. Head is greyer, with narrower moustache. Usually lacks rufous-buff wash on underparts. Females show heavier markings on belly and flanks.
[b] Juvenile. Head shows pale superciliaries and narrow moustache mark. Whitish underparts have narrow streaking, and whitish leg feathers have dark streaks.
[c] Adult male. Typical adult Peregrine from above. Note darker outer half of uppertail.

3. *F. p. brookei*. Darkest race. Averages smaller than two races above but larger than Barbary Falcon.
[a] Adult female. Overall dark and heavily marked. Note rufous on nape and dark rufous-buff wash on underparts.
[b] Juvenile male. Overall dark and heavily marked. Note thick dark streaking on underparts and chevron markings on leg feathers.
[c] Juvenile male. Dark brown upperparts have noticeable rufous feather edgings. Note dark head and wide dark moustache mark.
Juvenile female. is shown on Plate 41.

Similar species: Lanner Falcon, Eleonora's Falcon, Saker Falcon, Gyrfalcon

4. Barbary Falcon *Falco pelegrinoides*   p. 272; photos, p. 354

Smaller and paler than Peregrine. Desert-adapted. On perched falcons, wingtips reach or almost reach tail tip.

[a] Juvenile female. Buffy underparts show fine dark breast streaking. Note rufous-buff superciliary.
[b] Adult male. Adults have rufous napes and sparse dark markings on underparts.
[c] Adult male. Underwings are paler than those of other races. Note dark wrist comma and dark wingtips. Rufous nape is not always visible.
[d] Adult female. Upperparts of flying Barbarys appear much like those of adult Peregrines, except for rufous nape. Many adults appear paler due to sun fading.

Similar species: Lanner Falcon, Eleonora's Falcon, Saker Falcon

Plate 46

4[a] 4[b] 2[a] 1[a]

1[b]

3[a] 3[b]

2[b]

1[c]

3[c] 2[c]

4[c] 4[d]

Plate 47. MOROCCAN SPECIALITIES

1. Dark Chanting Goshawk *Melierax metabates*   p. 116

Chanting Goshawks have long legs and long tails. Adults have brightly coloured faces and legs. In flight they show long wings with rounded wingtips and long tail. Rare local resident in central Morocco. Vagrant to Israel.

[a] Juvenile. Overall brownish, with streaked breast and barred belly, leg feathers, undertail coverts, and underwings. Legs orangish.

[b] Adult. Head and breast dark grey, belly and underwing coverts finely barred grey, and undersides of flight feathers whitish. Note bright orangish-red face and legs.

[c] Adult. Dark grey above with variable amount of whitish marbling on uppersides of flight feathers and fine grey barring on white uppertail coverts.

[d] Juvenile. Dark brown above with fine dark barring on white uppertail coverts.

[e] Adult. Overall dark greyish; some lack whitish marbling on secondaries. Note bright orange-red cere, face skin, and long legs.

[f] Juvenile. Dark brown head and upperparts. Breast is streaked, and belly and flanks are barred. Note yellow face and cere and orangish legs.

[g] Adult. Overall dark greyish, with whitish marbling on secondaries. Note bright orange-red cere, face skin, and long legs.

Similar species: Distinct; no other Western Palearctic raptor similar

2. Tawny Eagle *Aquila rapax*   p. 177

Classic *Aquila* build. Juveniles are shown on Plates 32–34. Rare local resident in central Morocco.

[a] Adult. Overall medium brown to tawny-brown. Some adults lack dark breast streaks. Lacks wide dark band on tip of secondaries.

[b] Older immature. Often appear hooded: new darker feathers on head and breast contrast with faded juvenile feathers on belly. Eyes are brown.

[c] Adult. Typical adults have yellowish eyes and are overall tawny-brown with dark brown breast streaking. Wingtips reach tail tip. Note baggy 'trousers'.

Similar species: Steppe Eagle, Spanish Imperial Eagle, Eastern Imperial Eagle

3. Lappet-faced Vulture *Torgos tracheliotus*   p. 78; photos, p. 302

Race *negevensis* is shown on Plate 15. Race shown here is *tracheliotus*. Formerly bred in Morocco; now vagrant to north-western Africa.

[a] Adult. Overall dark blackish-brown with large red head, white bars on underwings, and white thighs.

[b] Adult. Overall dark blackish-brown with large red head and white thighs. Note large lappets on side of head. Some adults—males according to one authority—show some white feathers on back.

Similar species: Cinereous Vulture

Plate 47

1[a]

1[b]

1[c]

3[a]

1[d]

2[a]

1[g]

1[f]

1[e]

2[b]

2[c]

3[b]

Plate 48. ARABIAN SPECIALITIES

1. Bateleur *Terathopius ecaudatus* p. 88

Vagrant to Middle East. Unusual and distinctive short-tailed, long-winged, aerial raptor. 'Flying Wing'.

[a] Adult female. Adults have black body, red face, red legs that extend beyond tip of tail, and white underwing coverts. Females have white secondaries with black tips.

[b] Adult male. Adults have black body, red face, red legs that extend beyond tip of tail, and white underwing coverts. Males have black secondaries.

[c] Juvenile. Overall brown with tail longer than legs. Wings wider than those of adults. Sexes are alike.

[d] Subadult. Similar to juvenile but with narrower wings, pale secondaries with dark tips, and shorter tail (toes just reach tail tip). Sexes are alike.

[e] Juvenile head-on. Glides and soars with wings in a strong dihedral.

[f] Juvenile. Overall brown with yellow face skin, cere, and legs. Note wide cowl-like crest. Wingtips extend far beyond tail tip.

[g] Adult male. Adults have black heads and bodies with chestnut (rarely creamy-white) back and pale grey upperwing coverts. Note striking red face, cere, and legs. Short chestnut tail usually not visible, as it is covered by secondaries, which are black on males and white with black tips on females.

Similar species: None

2. Verreaux's Eagle *Aquila verreauxii* p. 202

Rare and local in the Middle East. Large black eagle. Wing shape of adult is distinctive.

[a] Adult head-on. Soars and glides with wings in a strong to medium dihedral.

[b] Adult male. Overall black with white 'U' on upper back and white uppertail coverts. These white areas are usually separated on males in flight but are connected on females.

[c] Older immature. Approximately 3-year-old eagle showing gradual change from juvenile to adult by slow replacement of juvenile with adult feathers.

[d] Adult. Overall black with white primary panels and yellow cere, lores, and toes.

[e] Juvenile. Appears somewhat dark but with whitish panels on primaries and whitish belly.

[f] Juvenile. Multicoloured plumage: blackish with a variable amount of white feather edgings, crown creamy-buff, and upper back chestnut. Wingtips reach tail tip.

[g] Adult. Overall black with white 'U' on back and white uppertail coverts. Wingtips reach tail tip.

[h] Juvenile. Multicoloured plumage: blackish with a variable amount of white feather edgings, creamy-buff crown, and white belly and undertail coverts.

Similar species: None

Plate 48

FAMILY AND SPECIES DESCRIPTIONS

Ospreys—Family Pandionidae

This family consists of only one species, the Osprey, which is a large, long-legged, eagle-like, fish-eating raptor. Ospreys are widespread throughout the northern hemisphere and Australia. Their outer toes, like those of owls, are reversible; this character and the sharp spicules on the lower surface of the toes help them to grasp slippery fish. Ospreys are anatomically different from raptors in Accipitridae, Cathartidae, and Falconidae, primarily in their specialized adaptations for capturing fish. Ospreys are thought to be most closely related to the kites in Accipitridae and are often included in that family in some taxonomic arrangements.

Osprey   *Pandion haliaetus*

Plate 5; Photos, p. 288

Description

A large, long-winged and long-legged raptor, usually found near water. In flight, its *gull-like crooked wings and white head with wide black eye-stripe* are distinctive. Sexes are almost alike in plumage, but females on average are somewhat larger than males. Cere and legs are dull blue-grey. Perched birds appear long-legged; their wingtips extend just beyond tail tip. Tail on distant flying birds can appear orangish.

Adult's white head has dark brown central crown and nape patches and *wide dark brown eye-lines* that extend down the sides of the neck to the shoulders. Iris is bright yellow. Back and upperwing and uppertail coverts are uniform dark brown. Underparts are white, usually with a more or less distinct band of short dark streaks forming a necklace across upper breast; females usually have more streaking, but there is much overlap and birds of either sex can be completely clear-breasted. Underwings show grey flight feathers, *black carpal patches*, and white underwing coverts with darker greater coverts forming blackish lines. Short tail appears dark with light bands from above, but light with dark bands from below.

Juvenile appears similar to adult, but back appears 'scaly' due to pale tips on most upperwing coverts and back feathers. Head is like that of adult except that it has narrow dark streaking in white areas of crown and nape. Iris is initially red and gradually fades to orange. Recently fledged birds show a rufous wash on napes and breasts, but this fades quickly and these areas are white by autumn migration. Flight feathers have wide pale tips. Undersides of secondaries are paler than those of adults. Tail has a wider white terminal band than that of adult.

Unusual plumages

Birds with a few white feathers in place of normally dark ones have been reported. Melanistic individuals have been reported from France and North America.

Similar species

Short-toed Snake Eagle (Plate 5) also shows white undersides, can show a similar dark necklace, and glides with wings held Osprey-like, but underwings lack the dark carpal patches and usually show bold black barring.

Bonelli's Eagle (Plate 36) juveniles in flight can appear somewhat similar with pale undersides, somewhat darker secondaries on underwing, and narrow dark band formed by greater underwing coverts, but they lack the dark carpal patches and have dark heads.

Large gulls can appear very Osprey-like but are smaller and have shorter, pointed wings, longer necks, and unbanded tails; they lack the black carpal patches on the underwings.

Flight

Powered flight is with slow, steady, shallow wingbeats with somewhat flexible wings. *Soars and glides with wings crooked in a gull-winged shape*, with wrists cocked forward and held above body level, and wingtips pointed down and back. Sometimes soars on flat wings. Hovers frequently while hunting over water.

Moult

Ospreys moult during the summer and winter and suspend moult during migration. Moult is typical of large raptors, with most primaries and all rectrices replaced annually. Secondaries are usually replaced every other year; juvenile and later secondaries are the same length. Moult of these feathers is asymmetric and random in older individuals. Juveniles begin moult on the winter grounds at an age of 5–7 months.

Behaviour

Ospreys are superb fishermen, catching prey with their feet after a spectacular feet-first dive, usually from a hover, but sometimes from a glide. They usually enter the water completely. They are able to take off from the water surface and, after becoming airborne, shake vigorously to remove water. They always carry captured fish head forward. Although they are almost exclusively fish-eaters, some non-fish prey items such as birds, turtles, and small mammals have been reported in their diet. They are usually found near water and are only away from it during migration.

Nest sites are usually near coasts and are most often in dead tree tops, but nests are also placed on man-made structures such as power poles, channel markers and other navigational aids, and even on small wooden docks. Ground nests have been found, most frequently on islands. Adults are vigorous in their nest defence, but seldom strike humans. Their high, clear, whistled alarm call, repeated continuously, is distinctive.

Courtship flights by males are a series of undulating dives and climbs, usually performed while carrying a fish and calling constantly. Sometimes flying adults swoop down and drag their feet through the water, a practice thought to be for cooling or cleaning their feet, but perhaps also a displacement behaviour.

There is much folklore about White-tailed Eagles robbing Ospreys of fish, and this happens occasionally; however, these two species cohabit water areas surprisingly peacefully, with only an occasional antagonistic encounter, initiated by either.

Status and distribution

Ospreys are found worldwide and are common near water in their main breeding areas of northern and eastern Europe; less common and local (formerly more numerous) in Scotland, Central Europe, south-western Portugal, and the western Mediterranean Sea. Breeding colonies in the Canary and Cape Verde Islands and the southern Sinai peninsula are of sedentary birds.

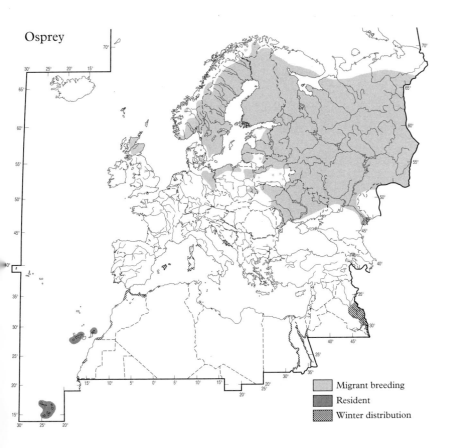

Osprey

Migrant breeding
Resident
Winter distribution

The entire northern population is migratory, leaving for the northern winter. Most move into Africa south of the Sahara; however, some birds remain in the Mediterranean.

All 1-year-olds remain south of the breeding grounds during their second summer. Two-year-old birds return north, some as far as the natal areas, but usually do not breed. A good number of 3-year-olds join the breeding population. Occasionally Ospreys are found far out at sea.

Fine points

Lack of a supraorbital ridge gives Ospreys a pigeon-headed appearance. In mated pairs, females almost always have the more distinct breast band, but there is overlap in this character. Clear-breasted, unmarked adult birds are usually, but not always, males.

Subspecies

The Western Palearctic subspecies is *Pandion h. haliaetus*. Four other subspecies occur in the Americas, South-east Asia, and Australia.

Etymology

Osprey comes from the Latin *ossifragus* meaning 'bone breaker', but this probably referred to the Bearded Vulture. *Pandion* after the mythical king of Athens, *haliaetus* from the Greek *hals* and *aetus*, for 'sea' and 'eagle'.

Measurements

Length: 53–66 cm (58 cm)
Wingspread: 147–174 cm (160 cm)
Weight: 1.0–2.0 kg (1.6 kg)

KITES—FAMILY ACCIPITRIDAE

The family Accipitridae consists of the kites, eagles, hawks, buzzards, vultures, and harriers.

Kites are thought to have evolved long ago and to be rather primitive raptors. Four species of kites occur regularly in the Western Palearctic; they belong to one of three subgroups: typical kites, white-tailed kites, and Honey Buzzards.

One Honey Buzzard, the Eurasian, occurs as a widespread summer breeder, and another, the Crested, occurs rarely as a vagrant on spring migration.

One white-tailed kite, the Black-shouldered, occurs as a local resident on the Iberian Peninsula, in north-western Africa, and along the Nile River in Egypt.

Two typical kites, Red and Black Kites, occur as widespread residents, but with some northern ones, especially Black Kites, migratory.

Eurasian Honey Buzzard *Pernis apivorus*
(Western Honey Buzzard/Honey Buzzard)

Plates 6, 7; Photos, p. 290

Description

A large buzzard-like kite, which is a widespread summer resident of forests throughout most of Europe. They lack bony projections over the eyes (supraorbital ridges) and *appear somewhat pigeon-headed. Small head and long neck and long tail with rounded corners* are distinctive on flying birds, which show an oval or rectangular dark carpal patch on each underwing. Wingtips fall short of tip of long tail on perched birds. When perched on the ground, they appear horizontal because of short legs. Plumages are polymorphic, but variations in base coloration of body and underwing coverts are similar for adult males, adult females, and juveniles. Age and sex are easily determined, however, using eye, cere, and face colour and extent of black on outer primaries. Base colour of body plumage is uniform and varies from white to creamy to pale brown to rufous to dark brown to black, with many intergrades in between these. Adults are marked on underparts with a variable amount of fine dark streaking, dark spotting, or wide dark barring or a combination of these; juveniles are marked only with wide dark streaking. Adults have yellow eyes and dark grey ceres; juveniles have medium brown eyes and yellow ceres. Females are only slightly larger than males.

Adult male. Sides of face are grey. Undersides of outer primaries show narrow black tips with a sharp line of contrast with rest of whitish feathers. Undersides of secondaries are whitish with wide dark terminal band and a wide unbanded space between it and the other bands. In good light, upperparts have a greyish cast and upperwings show dark trailing edges and dark tips to primary coverts. Underparts show a variable amount of dark spotting or narrow streaking on breast and dark barring on belly and flanks but can be completely unmarked. Tail above shows wide dark subterminal band and two narrower dark bands near base.

Adult female. Sides of face are brown; some paler-headed birds show dark lines behind eyes. Note small dark malar marks on sides of lower throat, noticeable on all but darkest birds. Upperparts are dark brown, with pale primary patches on upperwings, noticeable as pale primary patches on back-lighted wings. Undersides of outer primaries have dark tips that gradually shade into paler bases without sharp line of contrast. Undersides of secondaries have wide dark terminal band but usually lack the wide unbanded area of those of adult males. Underparts are almost always barred, on average more heavily than those of males. Tail above is either like those of males or with three dark bands near base.

Juvenile. Head is usually quite pale, sometimes completely white, but always with white forehead. Upperparts are dark brown, with pale primary panels on upperwings, pale tips to greater upperwing coverts, and pale uppertail coverts forming noticeable 'U'. Undersides of outer primaries are dark over more than outer half of feathers, with whitish bases. In fresh plumage, secondaries and inner primaries have wide pale tips. Undersides of secondaries are dusky, contrasting with whitish primaries and forming dark patches on underwings of flying birds. Greater secondary underwing coverts are paler than secondaries and other coverts, often noticeable as a buffy bar across underwings. Underparts are streaked. Tail above shows either four evenly spaced dark but less noticeable bands or an adult female pattern.

First summer birds are juveniles that have undergone some body moult; they have adult-like underparts and juvenile remiges and rectrices.

Unusual plumages

Albinism has been reported from the British Isles; partial albinism and dilute plumages have been reported in Italy.

Similar species

Common Buzzards (Plates 24–7) in most plumages appear similar but have shorter necks and thicker heads, shorter, more square-cornered tails, more pointed wingtips, and different tail patterns. In addition, their active flight is with shallower wingbeats of stiffer wings. Adults and many juveniles show a pale 'U' between breast and belly, a feature never seen on Eurasian Honey Buzzards. Dark juvenile Buzzards lack pale bands on underwings characteristic of juvenile Eurasian Honey Buzzards. Dark carpal patch on underwings of Common Buzzard is square rather than oval or rectangular like that of Eurasian Honey Buzzard.

Black Kite (Plates 8, 9) juveniles can appear similar to juvenile rufous Eurasian Honey Buzzards, with dark secondaries, pale primary panels, and pale lines between secondaries and coverts, but head and neck are shorter, tail tip is forked, and wings are held with wrists up.

Flight

Powered flight is with slow, deep wingbeats of somewhat elastic wings. Soars on flat wings, occasionally with a small dihedral; glides with wrists level but wingtips noticeably drooped. They never hover.

In flight they often twist their heads from side to side in a snake-like manner and twist their tails in a kite-like fashion while manoeuvring.

Moult

Annual moult of adults is complete; it begins in late summer but is suspended during migration and completed on African winter grounds. Juveniles replace some body feathers and coverts over winter, but begin replacing remiges and rectrices in summer.

Behaviour

Eurasian Honey Buzzards are specialized kites that feed mainly on larvae, pupae, nests, and adults of various wasps and bees. But they also feed at times on other insects, reptiles and amphibians, small mammals, and eggs and nestlings of birds, as well as some fruit and berries. They hunt from low perches, searching for movements of insects to and from nests, and have been reported to follow flying insects back to their nests. They are comfortable walking around on the forest floor.

They make stick nests like buzzards and hawks and their display flights are similar.

They are shy and retiring and are not aerial except for display flights and migration, as a result, they are not often seen.

On migration they form communal night roosts, but apparently do not do this while breeding in Europe. While migrating, they are more likely to continue flying using active flight in absence of thermals than are buzzards.

Status and distribution

Eurasian Honey Buzzards are fairly common summer-breeding residents throughout most of the forested areas of Europe, absent only from tundra and treeless steppe. They are rare in the British Isles.

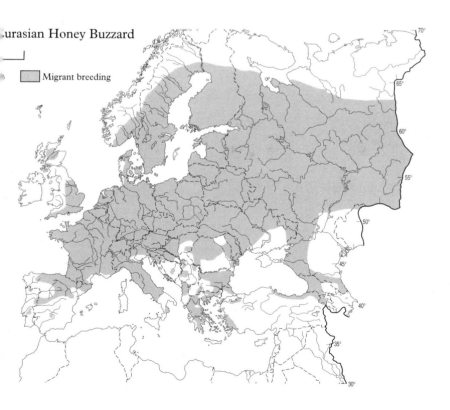

urasian Honey Buzzard

Migrant breeding

Entire population is migratory, with large numbers being recorded at Gibraltar, the Bosphorus, and in Israel on migration. Main wintering area is west and central equatorial Africa, but they are recorded also in southern and eastern Africa. Two were reported overwintering in southern Italy recently.

Some references state that juveniles remain on the African winter grounds, and no doubt many do so, but juveniles are seen regularly on spring migration heading north and there are many records in summer from southern Europe and a few from northern Europe.

Fine points

Eurasian Honey Buzzards have rather feeble beaks and straight talons, adaptations for eating insects and digging up insect nests.

Subspecies

None. Some authorities consider the Crested Honey Buzzard *Pernis ptilorynchus* of Asia to be conspecific, but its plumages are always distinct (especially lacking dark carpal patches), and there are no known intergrades.

Etymology

Common name was coined by Willoughby, who, in the 1600s, found combs of wasps in the nest. *Pernis* is thought to be a corruption of Greek *pternis* for 'a bird of prey'; *apivorus* comes from Latin *apis* and *voro* for 'bee' and 'devour', indicating its food preference.

Measurements

Length: 51–57 cm (54 cm)
Wingspread 115–136 cm (130 cm)
Weight: 510–1050 g (790 g)

Crested Honey Buzzard   *Pernis ptilorynchus*
(Oriental Honey Buzzard)

Plates 6, 7

Description

A large buzzard-like kite, which is a rare vagrant to the Middle East and Turkey from eastern Asia. They lack bony projections over eyes (supraorbital ridges) and *appear somewhat pigeon-headed. Small head and long neck and long tail with rounded corners are distinctive* on flying birds. Compared to Eurasian Honey Buzzard, Crested Honey Buzzards have broader wings, shorter tails, and lack dark carpal patches on underwings. Short crest is seldom noticeable. Wingtips fall short of tip of long tail on perched birds. When perched on ground, they appear horizontal because of short legs. Plumages are polymorphic, but variations in base colorations of body and underwing coverts are identical on adult males, adult females, and juveniles; age and sex are easily determined, however, using eye, cere, and face colour and amount of black on outer primaries. Base colour of body plumage is uniform and varies from white to creamy to pale brown to rufous to dark brown to black, with many intergrades in between these. Adults are marked on underparts with a variable amount of fine dark streaking, dark spotting, or wide dark barring or combination of these; juveniles are marked only with wide dark streaking. Adults have dark grey ceres; juveniles have yellow ceres. Females are only slightly larger than males.

Adult male. Eyes are dark brown, and sides of face are grey. Undersides of outer primaries show narrow black tips with a sharp line of contrast with rest of whitish feathers. Undersides of secondaries are whitish with wide dark terminal band. In good light, upperparts have a greyish cast and upperwings show dark trailing edges and dark tips to primary coverts. Underparts show a variable amount of dark spotting or narrow streaking on breast and dark barring on belly and flanks but can be completely unmarked. Tail is dark brown with a wide pale band across its centre.

Adult female. Eyes are yellow, and sides of face are brown; paler-headed birds show dark lines behind eyes. Upperparts are dark brown, usually with pale primary patches on upperwings. Undersides of outer primaries have dark tips that gradually shade into paler bases lacking sharp line of contrast. Underparts are almost always barred, averaging heavier than that of males. Tail above is pale with three dark bands near base.

Juvenile. Head is usually quite pale, sometimes completely white, but always with white forehead. Upperparts are dark brown, with pale primary panels on upperwings, pale tips to greater upperwing coverts, and pale uppertail coverts forming noticeable 'U.' Undersides of outer primaries are dark over more than outer half of feathers, with whitish bases. In fresh plumage, secondaries and inner primaries have wide pale tips. Undersides of secondaries are dusky, contrasting with whitish primaries and forming dark patches on underwings of flying birds. Underparts are streaked. Tail above shows either four evenly spaced narrow dark bands or a pattern like that of adult female. Eyes are brown.

First Summer birds are juveniles that have undergone some body moult; they have adult-like underparts and juvenile remiges and rectrices.

Unusual plumages

No unusual plumages have been reported.

Similar species

Eurasian Honey Buzzards (Plates 6, 7) in most plumages appear similar but have narrower wings (showing five, not six, 'fingers' on wingtips), longer tails, and almost always show black carpal patches (never on Crested Honey Buzzard). Adult males have yellow eyes and pale tail with narrow dark bands.

Flight

Powered flight is with slow, deep wingbeats of somewhat elastic wings. Soars on flat wings, occasionally with a small dihedral; glides with wrists level but wingtips noticeably drooped. They never hover.

In flight they often twist their heads from side to side in a snake-like manner and twist their tails in a kite-like fashion while manoeuvring.

Moult
Annual moult of adults is complete and begins in summer but is suspended during migration and completed on winter grounds. Juveniles replace some body feathers and coverts over winter, but begin replacing remiges and rectrices in summer.

Behaviour
Crested Honey Buzzards are specialized kites that feed mainly on larvae, pupae, nests, and adults of various wasps and bees. But they also feed at times on other insects, reptiles and amphibians, small mammals, and eggs and nestlings of birds, as well as some fruit and berries. They hunt from low perches, searching for movements of insects to and from nests, and have been reported to follow flying insects back to their nests. They are comfortable walking around on the forest floor.

They make stick nests like buzzards and hawks and their display flights are similar.

They are shy and retiring and are not aerial except for display fights and migration, as a result, they are not often seen.

Status and distribution
Occurs in the Western Palearctic as a rare vagrant, with a few recent records from Turkey, Egypt, Israel, Saudi Arabia, and southern Italy.

Entire north-eastern population is migratory; the bulk moving south into south-eastern Asia. Indian subcontinent population is sedentary, with possibly some southward movement in the northern winter.

Fine points
Many Crested Honey Buzzards have a distinct dark collar across their upper breasts, a feature not seen on Eurasian Honey Buzzard.

Individuals of the more migratory northern race, *orientalis*, the one most likely in the Western Palearctic, often lack the short crest.

Subspecies

Orientalis occurs in north-eastern Asia south- to north-eastern China, Korea, and Japan; *ruficollis*, on the Indian subcontinent. Some authorities consider this species a race of the Eurasian Honey Buzzard *Pernis apivorus*, but their plumages, particularly those of adult males, are always distinct, and there are no known intergrades.

Etymology

Common name was coined by Willoughby, who found combs of wasps in the nest in the 1600s. *Pernis* is thought to be a corruption of Greek *pternis* for 'a bird of prey'; *ptilorynchus* comes from Greek *ptilon* and *rhunklhos* or 'feather' and 'bill', indicative of its dense feathers on forehead and lores, a protection from bees and wasps.

Measurements

Length: 52–68 cm (60 cm)
Wingspread: 130–150 cm (145 cm)
Weight: 750–1500 g (1190 g)

Black-shouldered Kite   *Elanus caeruleus*
(Black-winged Kite/Common Black-shouldered Kite)

Plate 20; Photos, p. 292

Description

This kite of North Africa and south-western Iberia is gull-like in coloration (grey, white, and black), owl-like in silhouette (large head, wide wings, and short tail) and flight (soft, steady wingbeats), and falcon-like in wing shape. Its distinctive behaviour, outline, and coloration facilitate identification. *Black shoulder* of perched birds is formed by charcoal black median and lesser secondary upperwing coverts. This also shows as a distinctive *black area on upperwing* on flying birds. Sexes are almost alike in size and plumage. Juvenile plumage is similar to adult's. Cere is pale yellow, bill is black, and legs are yellow-orange. Wingtips extend far beyond tail tip on perched birds; tail tips appear slightly forked on folded tails.

Adult's large head is mostly white except for medium grey crown and nape and small black areas in front of eyes that extend as lines above and behind eyes. Back, greater secondary and primary upperwing coverts, and uppertail coverts are medium to dark grey (males average paler). Uppersides of flight feathers are dark grey. Underparts are uniformly white. Underwings consist of dark grey primaries and outer secondaries, paler grey inner secondaries, and white coverts. *Short tail* is white, except for pale grey central feathers. Iris varies from orange-red to scarlet.

Juvenile is similar to adult, but crown and nape are greyish-brown with narrow whitish streaking. Iris colour varies with age, going from light brown to yellow to orange. Grey-brown back feathers and greater upperwing coverts have white edging. Underparts are white with some narrow black shaft streaking and with a rufous wash across breast that fades shortly after fledging. Flight feathers have white tips. Tail appears completely grey from above. (Most juveniles moult into adult plumage soon after independence; see 'Moult'.)

Unusual plumages

No unusual plumages have been reported.

Similar species

Pallid Harrier (Plate 20) (and Montagu's and Hen Harrier) adult males are similar in coloration and shape, but have longer tails, narrower wings, and different pattern of black on primaries and lack black upperwing coverts.

Kestrels (and Lesser Kestrels) (Plates 38, 39) while hovering can appear similar but have longer tails and brownish, not greyish, upperparts.

Flight

Powered flight is owl-like with soft, deliberate wingbeats. Soars and glides with wings in a strong dihedral. Sometimes glides with wings in a modified dihedral. Hovers frequently, often with legs extended.

Moult

Moult is not well known but apparently is completed annually although usually suspended while breeding and during winter. Juveniles begin post-juvenile moult of body feathers and wing coverts shortly after fledging.

Behaviour

Black-shouldered Kites hunt either from exposed perches or by hovering; hunting mostly in early morning and late afternoon. They prey on small mammals, birds, insects, and reptiles. While perched, they often droop their wings below the tail and pump the tail up over the back and down. This is thought to be an aggressive signal to other kites. They are persistent in hovering, often dangling their feet while doing so. When prey is spotted, they slowly descend with the wings held in a high 'V' above the body and finish with a rapid stoop with feet and talons outstretched.

They are strongly territorial when nesting. Display flight of adult male is with fluttering wings held in a 'V' above the body and is accompanied by calls. Agonistic interactions between neighbours are frequent. Non-breeding birds form communal night roosts, mainly in fall and winter, in Africa and Asia, but this behaviour has not been reported from the Western Palearctic.

Status and distribution

Sedentary residents of lightly wooded grasslands and croplands of west-central Spain (Extramadura), southern Portugal, Mediterranean coasts of Morocco and Algeria, central Morocco, and along the Nile River in Egypt, and recently in south-western France. Vagrants are occasionally reported in Europe as far north as Poland and Denmark and as far east as Greece and in the Middle East.

Fine points

Adult's central tail feathers are grey; its others are white. Folded tail appears grey above and white below.

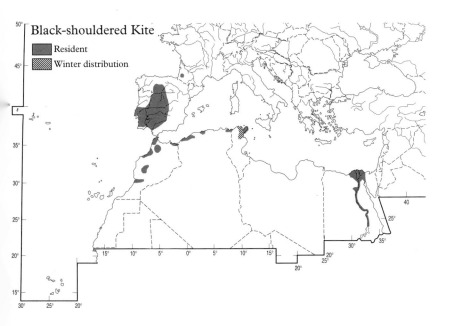

Subspecies

E. c. caeruleus occurs in the Western Palearctic.

Etymology

Elanus is Latin for 'a kite'; *caeruleus* is Latin for 'sky blue', its back colour. Also called 'Black-winged Kite'.

Measurements

Length: 29–33 cm (31 cm)
Wingspread: 80–88 cm (85 cm)
Weight: 180–250 g (220 g)

Red Kite   *Milvus milvus*

Plate 8; Photos, p. 294

Description

The region's largest kite breeds only in the Western Palearctic. Its rufous coloration, buoyant flight, and flight silhouette are distinctive. It could only be confused with Black Kite. *Large squarish white primary panels on underwings are diagnostic.* It is one of many raptors that have a pale bar across each upperwing. *Wingtips barely reach tip of shortest (central) tail feathers on perched birds, but far short of tail tip. Tail is deeply forked*, with outer feathers much longer than central ones, and appears forked except when fully spread. Sexes are alike in plumage; females are noticeably larger. Juvenile plumage is different from that of adult.

Adult head and neck are whitish with narrow black shaft streaking on face and heavier, wider black streaking on crown and nape. Back is dark brown with most feathers having rufous-buff edges. Upperwing coverts are dark brown with wider rufous–buff edges; paler median secondary coverts form pale bar across each upperwing. Uppersides of flight feathers are dark brown. Uppertail and greater uppertail coverts are bright rufous; outer tail feathers have blackish tips. Underparts are rufous with wide black shaft streaking on breast and belly but not on leg feathers and undertail coverts. Underwings show a strong contrasting pattern of white primaries, with black tips on outer ones, dark brown secondaries, blackish greater primary and secondary coverts, and rufous median and lesser secondary coverts. Pale undertail contrasts with rufous undertail coverts. Iris is yellow; cere and legs are orangish-yellow. Beak is black, with variable amount of orangish-yellow at base.

Juvenile. Buffy head has black shaft streaking on crown and nape. Back and upperwing coverts are like those of adult, except that edging is more rufous-brown; pale upperwing bars may be more noticeable than those of adult. Dark brown flight feathers have pale tips. Uppertail is rufous-brown, less reddish in appearance than those of adults. Underparts appear different: breast is brown with wide buffy shaft streaking; belly, leg feathers, and undertail coverts are buffy and unstreaked. Latter do not contrast with pale undertail. Underwing pattern is similar to that of adult, but inner primaries and secondaries have dusky tips that form a band across trailing edge of primaries. Greater upper- and underwing coverts have narrow pale tips in fresh plumage; these form a narrow band across wings until tips wear off. Iris is greyish in nestlings, then becomes brown, and later, pale yellow on older birds; cere and legs are yellow. Beak is black.

Unusual plumages

Hybrids with Black Kites have been reported. Juveniles have intermediate characters. A dilute-plumage juvenile was seen in Italy.

Similar species

Black Kite (Plates 8, 9) has similar silhouette and flight style. See under that species for distinctions.

Flight

Superb flyers, effortlessly using whatever wind is available to remain airborne, often without a wing flap, but with constant adjustments of wings and tail. Powered flight is with deep slow wing beats of flexible wings. Soars with wings somewhat arched; glides with wings more arched.

Moult

Annual moult is complete. Post-juvenile moult may be only of body feathers; it is not well known.

Behaviour

Red Kites are elegant on the wing and spend much of their time in foraging flight. They are more often found near woodlands and are less gregarious than are Black Kites. However, small groups will gather at abundant food sources. Like Black Kites, they are opportunistic and take a wide spectrum of food by predation, scavenging, and piracy. They regularly take waste from slaughterhouses, as well as offal and all manner of material from rubbish dumps. Food is often snatched from the ground without alighting after a short, steep stoop, after which the kite rises rapidly.

They usually nest singly, building their nests in trees and defending the immediate nest area.

Red Kites form small communal night roosts during the non-breeding season and migrate in small groups (which may be family groups). They join Black Kite winter night roosts in some areas.

Status and distribution

Resident throughout most of the Iberian peninsula and southern Italy, including Sicily, as well as the Balearic Islands, Corsica and

Red Kite

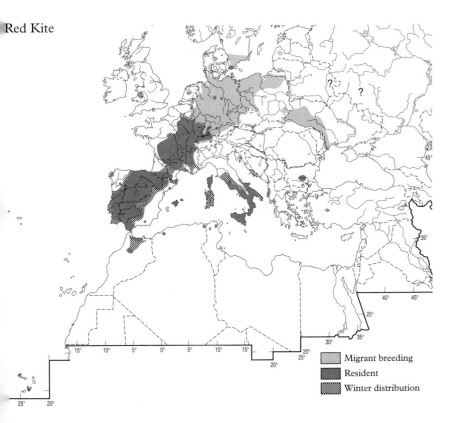

Migrant breeding

Resident

Winter distribution

Sardinia. They occur locally in northern Morocco, central and north-eastern France, Wales, parts of the former Yugoslavia, and the eastern Black Sea. Farther north, they occur from Belgium and Holland through northern Germany and Poland and east to Russia. Northern continental kites are migratory, moving south into southern Europe and Africa for winter. A few winter in the Middle East.

They are being reintroduced successfully in Scotland and England.

Fine points
None.

Subspecies
Milvus m. milvus occurs throughout, except for *M. m. fasciicauda* on the Cape Verde Islands. They are similar to nominate birds but are smaller and browner with less forked tails. Some authorities think that these are hybrids with *M. migrans*.

Etymology
Milvus is Latin for 'kite'.

Measurements
Length: 60–66 cm (63 cm)
Wingspread: 154–170 cm (162 cm)
Weight: 750–1600 g (1075 g)

Black Kite   *Milvus migrans*

Plates 8, 9; Photos, p. 296

Description

A large dark brown kite, a widespread and common breeder throughout the Western Palearctic except for the British Isles, Scandinavia, the Middle East, and north Africa, but local along the Nile River in Egypt and in north-western Africa. Its buoyant flight and flight silhouette are distinctive; it could only be confused with Red Kite, dark-morph Booted Eagle, or all-dark Marsh Harrier. Most birds show whitish crescent-shaped to squarish primary panel on each otherwise dark underwing; dark barring in panels noticeable on close birds. It is one of many raptors that have a pale bar on each upperwing. *Wingtips reach beyond central tail feathers to near tail tip on perched birds.* Tail is forked with outer tail feathers somewhat longer than central ones but appears square when fully fanned; occasional birds (usually juveniles) have rounded tail tips, lacking any fork. Sexes are alike in plumage; females are noticeably larger. Juvenile plumage is different from that of adult. Cere and legs are yellow. Beak is black.

Adult face and crown are greyish with heavy narrow black shaft streaks, but nape is dark brown. Upperparts are dark brown with paler median secondary coverts forming pale wing bars on upperwings. Uppersides of flight feathers are dark brown. Underparts vary from sooty-brown to dark brown, with narrow black shaft streaks on breast and sooty-rufous wash on belly, leg feathers, and undertail coverts. Tail above is dark reddish-brown with darker brown banding but appears whitish below with dark brown bands. Iris is yellow. *Note*: a few adults are overall dark brown or blackish.

Juveniles are similar to adults but usually lack sooty-rufous tones on underparts. They have pale brown heads with a dark brown shaft streaks and a dark brown rectangular area behind each eye, wide pale shaft streaks on breast, and whitish feather edges to greater wing coverts; these form narrow pale lines across middle of upper- and underwings when plumage is fresh. Back and other wing coverts also have pale edges; upperparts appear more speckled than those of adult. Pale primary underwing panels are usually larger than those of adults. Belly, leg feathers, and undertail coverts are usually paler than breast. Iris is brown but becomes pale yellow by spring. Note: a few juveniles are overall dark brown (see Plate 9).

[*Note*: Some adults seen in the Middle East appear decidedly more rufous overall with large pale squarish primary panels (Plate 9). Resident adults along Nile River have yellow beaks (Plate 9).]

Unusual plumages

A pure white Black Kite was sighted at Gibraltar. Hybrids with Red Kites have been reported that have intermediate characters.

Similar species

Red Kite (Plate 8) has similar silhouette including forked tail but is larger, more rufous overall, longer-winged and -tailed, and appears overall more slender. Adult's head is completely whitish. Wing panels on underwings are larger and whiter. Wingtips do not reach shortest tail feathers on perched birds. Tail is more deeply forked.

Marsh Harrier (Plates 16, 17) in flight can appear similar, even to having pale underwing panels (adult female), but lacks forked tail and pale bars on upperwing and holds wings straight, not arched.

Booted Eagle (Plate 37) dark or rufous morph soaring or gliding can appear similar, even having pale median upperwing bars and similarly shaped arched wings, but shows white 'U' on uppertail coverts, pale panels restricted to inner primaries, and white 'headlights' when viewed head on, and lacks forked tail. Wrist is relatively further from body on Booteds, that is, their arms are longer.

Flight

Graceful flight is usually with few wingbeats but with constant minor adjustments of wings and twisting of tail. Powered flight is with slow, almost floppy wingbeats of somewhat flexible wings. Soars with wings somewhat arched; glides with wings more arched.

Moult

Annual moult is usually complete but is suspended for migrations. Juveniles begin post-juvenile moult in May–June, same time as adults.

Behaviour

The gregarious Black Kites are wonderful flyers, often remaining airborne for hours without a wingbeat. They are usually found near water. Large groups gather where food is readily available (e.g. rubbish dumps, insect swarms, mammal outbreaks). They take a wide array of food, including birds, mammals, insects, and reptiles, especially fish, but also garbage, refuse, and offal. They will chase each other or smaller raptors, or even other birds, trying to steal a morsel. Food is often snatched from the ground or water surface without alighting after a short steep dive, after which the bird rises rapidly.

They are somewhat vociferous, particularly when interacting with each other. Their drawn out tremulous call can be heard for some distance.

They prefer to built their nests in trees, usually near water bodies. They nest singly or in loose colonies of up to 30 pairs.

They form communal night roosts of up to thousands, especially in winter, gathering at roost sites in the afternoon and soaring about, sometimes going up out of sight. Most come into the roost before dark, but some birds do not enter until long after dark.

Status and distribution

Fairly common but local summer breeders throughout most of Europe, except for the British Isles, Scandinavia, most of northern Africa, and the Middle East. In north Africa they are locally common in the north-west and along the Nile River of Egypt.

Most birds leave Europe for winter, migrating into southern Africa, but some stay in North Africa and the Middle East.

Fine points

Dark kites can appear much like dark juvenile Marsh Harriers. They are distinguished by wing proportions and attitude in flight. Some

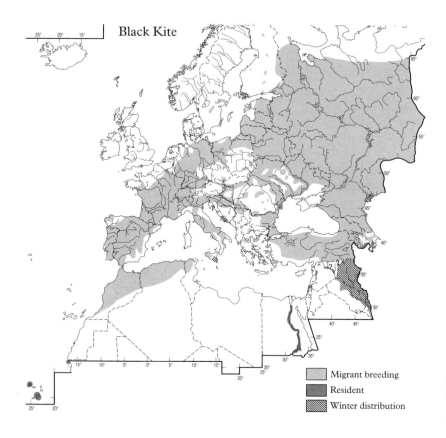

Black Kite

Migrant breeding

Resident

Winter distribution

adults seen in the Middle East have very rufous underparts and resemble Red Kites, but they have buffy, not white wing panels and usually lack the Red Kite's black line through underwings.

Subspecies

Milvus m. migrans occurs throughout Europe and locally in north-west Africa. *M. m. aegyptius* occurs along the Nile River in Egypt. Asian birds migrating through the Middle East are thought to be *M. m. lineatus*.

Etymology
Milvus is Latin for 'kite'; *migrans* is Latin for 'migrating'.

Measurements
Length: 46–59 cm (53 cm)
Wingspread: 130–155 cm (143 cm)
Weight: 560–1210 g (875 g)

Sea and Fishing Eagles—genus *Haliaeetus*

Sea and Fishing eagles of the genus *Haliaeetus* are large eagles usually found near water, where they can prey on their favourite food, fish. They also take a variety of other prey, including waterfowl, regularly eat carrion, and pirate prey from other raptors.

They are characterized by large size, large beaks, pale heads in adult plumage, and unfeathered lower tarsi.

Only one species, the White-tailed Eagle, breeds in the Western Palearctic. Two others, Bald Eagle and Pallas's Fish Eagle, occur as rare vagrants.

Haliaeetus is from the Greek *halos* and *aetos* for 'sea' and 'eagle'.

Pallas's Fish Eagle   *Haliaeetus leucorhyphus*

Plate 11

Description

A rare vagrant to the Western Palearctic from central Asia, this is a large dark eagle that is somewhat smaller and narrower-winged than White-tailed Eagle. Adult plumage is reached in 4 or 5 years. Sexes are alike in plumage; females are larger. Wingtips fall just short of the tail tip on perched eagles. Legs are dusky yellow.

Adults have off-white heads and necks and dark brown flight feathers and coverts. Upper back and underparts are warm dark reddish-brown. Black tail shows a *wide white band across its centre*. Eyes are light grey-brown and beak and cere are dull grey.

Juveniles are overall paler brown except for dark brown ear patches, flight feathers, and tail and a white bar across each underwing that, with the white inner 3 to 5 primaries, form a pale 'M' on the underwings, especially noticeable on distant eagles. Back and upperwing coverts are somewhat paler than flight feathers, resulting in two-toned uppersides. Underparts often fade to creamy or dirty white. Many moult head feathers giving them a dark hooded appearance. Eyes are dark brown, and beak and cere are dark.

Second winter eagles are similar to juveniles but appear overall more mottled due to new darker feathers and new dappled leg feathers and undertail coverts. New dark tail shows faint to fairly conspicuous wide pale mottled band across its centre.

Third winter eagles are similar to younger eagles but appear even more mottled, with pale 'M' on underwings less distinct and wide pale band across centre of tail more distinct.

Fourth winter eagles are in transition to adult plumage. Head is mostly white, often with a trace of the dark ear patches, and dark tail has wide white band with some dark spotting across its centre.

Unusual plumages

No unusual plumages have been recorded.

Similar species

White-tailed Eagles (Plates 10, 11) appear somewhat similar in all plumages but are larger and bulkier and have wider wings. They show seven 'fingers' on wingtips; Pallas's show only six. Adults have all white tails; immatures have white in the tail and lack the pale 'M' on undersides.

Eastern Imperial Eagle adults (Plate 34) are similar in shape in flight with long heads and necks and parallel leading and trailing edges of wings but are overall a darker blackish-brown, lacking any reddish tones, and have dark throats, different tail pattern, and pale undertail coverts.

Flight

Active flight is with rather rapid wingbeats, rather more agile compared to that of White-tailed Eagle. Soars with wings held level. Glides with wings somewhat cupped, with wrists cocked up and wing tips down. Pallas's Fish Eagles soar often.

Moult

They undergo moult from spring through to autumn, usually suspending it in winter. Replacement of flight and other feathers is not complete annually. Post-juvenile moult begins in March of April of the second calendar year; however, many have replaced head feathers by their first winter and appear 'dark hooded'.

Behaviour

Usually found near water where they can find their favourite food, fish. Breeding areas are almost always near water. In winter, when fish may be scarce, they also eat carrion and waterfowl. They are superb fishermen but also agile aerial predators that regularly capture waterfowl, and sometimes pirate food from other raptors.

Status and distribution

Recorded as a vagrant in Finland, Norway, Poland, and Holland. Formerly occurred regularly along the Caspian Sea coast of the Caucasus in winter, where now recorded occasionally, and in Iraq, where there are no recent records. Three recent records from Israel are being considered.

They are local in areas with small and large bodies of water throughout central and southern Asia, including northern India and Pakistan, where they breed in the winter and migrate north in summer.

Fine points

Immature Pallas's Fish Eagles differ from immature White-tailed Eagles by dark ear patches, pale 'M' on underwings and, only six 'fingers' on wing tips.

Subspecies

Monotypic.

Etymology

'Pallas' after Peter Simon Pallas (1741–1811), a German-born zoologist who collected the first specimen of this eagle on the lower Ural River near the Caspian Sea; *leucoryphus* is from the Greek *leukos* for 'white' and *koruphos* for 'crowned', or 'white-crowned'.

Measurements

Length: 72–85 cm (79 cm)
Wingspread: 200–230 cm (215 cm)
Weight: 2.4–3.6 kg (3.9 kg)

White-tailed Eagle   *(Haliaeetus albicilla)*
(White-tailed Sea Eagle)

Plates 10, 11; Photos, p. 298

Description

A large sea eagle, an uncommon to rare local breeding resident throughout north-eastern Europe, coastal Scandinavia, the coastal Baltic, and locally throughout eastern Europe and Turkey but more widespread in winter; usually found near water. Sexes are alike in plumage; females are larger. In all plumages distinguished by rather straight leading and trailing edges of long, broad wings; short, wedge-shaped tail; long head and neck; and unfeathered tarsi. Adult plumage is attained after four or five annual moults.

Adult head and neck vary from light brown to buff to cream, paler on some (older?) adults, with a variable amount of narrow brown streaking and usually with a narrow dark eye-line. Body and coverts are dark brown with a variable amount of buffy feather edging that is wider, thus appearing paler, on breast, upper back, and upperwing coverts. Flight feathers, leg feathers, and undertail coverts are dark brown. Short white tail has wedge-shaped tip; some adults show black spots on a tail feather or two throughout their lives. Wingtips reach or almost reach tail tip on perched eagles. Beak and cere are lemon yellow, and eyes are yellow.

Juvenile head, body, and wing and tail coverts are dark brown with an overall uniform coloration. Some juveniles show whitish and blackish streaking on breast, have paler brown to buffy backs, bellies, and upperwing coverts, and can show buffy leg feathers. The only white on juveniles' bodies and wings are the axillaries and usually a white bar across each underwing (median secondary coverts); these are features shared by the first three plumages. Long tail shows dark outer webs and (usually) white inner webs with triangular dark tips resulting in an arrow-shaped white area on the tip of each feather (feature shown on first three plumages). Trailing edge of wings appears serrated due to pointed tips of secondaries and is more curved than that of adults. Wingtips fall somewhat short of tail tip on perched juveniles. Beak and cere are blackish, and eye is dark brown.

Older immatures appear more mottled than juveniles, as new body and covert feathers are often a mixture of white and dark brown and retained feathers are faded. Assigning eagles to age classes other than juvenile and adult in the field is not easy but is possible with good looks at close range.

Second winter eagles appear somewhat like juveniles but are overall more mottled due to mix of old faded and new brown and white feathers; whitish areas usually evident on upper back and belly. New secondaries are shorter, darker, wider, with blunt tips. Usually from three to six inner primaries and a few secondaries (usually S1, S5, and S13, but as many as S1–2, S5–6, and S11–13) are replaced after the first moult; trailing edges of wings appear uneven and ragged as a result. Beak and cere are still dark but with a narrow pale line at base of beak. Eyes are medium brown.

Third winter eagles are similar to younger eagles but now show some dull yellow areas on beak and cere and more mottled dark feathers in plumage. Most secondaries have been replaced, but juvenile S4, S8, or S9 or combination thereof may be evident as longer pointed feathers on the trailing edge of wings. Tail in this plumage is highly variable, from white with dark tips to typical immature.

Fourth winter eagles are similar to younger eagles but now have dull yellow beaks and light brown eyes. New feathers on head, neck, body and wing coverts are grey-brown with pale feather edges. Wing linings are now mostly dark, with some white spotting. Tail is variable from typical immature to white with dark band on tip.

Fifth winter eagles are in their first adult plumage but may be distinguished by retained immature feathers, by black tips on some tail feathers, or by dusky culmen or brownish tint to eyes or both. Younger adults apparently have darker heads and necks, compared to older adults.

Unusual plumages

Both albinism and partial albinism have been reported. A dilute-plumage adult specimen from Ireland was overall greyish-white.

Love (1983, p. 13) describes adult specimens from Scotland in similar plumage.

Similar species

Eastern Imperial Eagle (Plate 34) adults and older immatures can appear similar to immature White-tailed Eagles in flight. See under that species for distinctions.

Greater Spotted Eagle (Plates 29, 30) can appear similar to immature White-tailed Eagles in flight but are much smaller, have white Aquila patches at bases of upper primaries, and lack white on axillaries and white in the tail.

Flight

Large, heavy, often cumbersome predators but at times surprisingly agile in flight. Powered flight is with heavy, but shallow and rather quick wingbeats. Soars with wings held level or in a slight dihedral. Glides with wings held level or slightly depressed.

Moult

They undergo an incomplete annual moult, actively moulting from March into November. They only replace some of their flight, tail, covert, and body feathers each year. Adults may also suspend moult for a time while raising young. Juveniles begin their first moult in late spring and moult in the typical accipitrid pattern (see 'Moult' in the Introduction). Adult plumage is attained after four or five moults.

Behaviour

Prey mainly on both fish and birds, primarily ducks and seabirds, but also take mammals, live and as carrion. They hunt both on the wing and from perches, as well as by wading in shallow water along shores or in rivers and streams. Fish are usually snatched from the surface of the water. Birds, often crippled or sick ones, are taken as they take off from the water or ground or as they surface from a dive, less often in the air. Some nestling seabirds are taken from the nest.

Flight displays include undulating flight and mutual soaring, usually accompanied by loud vocalizations. Typical large raptor stick

White-tailed Eagle

Migrant breeding
Resident
Winter distribution

nests are built on large cliff faces or high up in a large tree just below the canopy, but occasionally on the ground. One to three eaglets are usually raised.

Status and distribution

Uncommon breeding residents of coastal Norway, much of the Baltic Sea coast, and the Volga River delta and rare and local breeding residents near water bodies in eastern Iceland, some Scottish Islands (where reintroduced), Lapland, much of eastern Europe, and Turkey. Eagles from the north-eastern part of the range migrate south for the winter. Immatures disperse away from breeding areas, particularly in winter.

Their populations have been reduced in many areas due to direct human persecution (shooting, poisoning, and nest destruction) and electrocution.

Conservation activities in the Baltic region, including winter feeding stations, have resulted in population increases in recent years.

Fine points
Head and long neck stick out further on White-tailed Eagles, compared to other large eagles, projecting almost as far from the body as does the tail of adults.

Subspecies
Eurasian race is *H. a. albicilla*; Greenland race is *H. a. groenlandicus*.

Etymology
albicilla is from Latin *albus*, 'white', and *illus*, a Latin diminutive suffix, used mistakenly to mean 'tailed'. The error probably originated from the name of the wagtail genera, *Motacilla*.

Measurements
Length: 77–95 cm (86 cm)
Wingspread: 199–250 cm (226 cm)
Weight: 3.0–6.8 kg (4.8 kg)

Bald Eagle   *Haliaeetus leucocephalus*

Plate 11

Description

A rare vagrant to the British Isles from North America, this is a large sea eagle that is similar in all plumages to the White-tailed Eagle. Five age-related plumages can be recognized in the field. Adult plumage is reached in 4 or 5 years. Sexes are alike in plumage; females are larger. Wingtips reach or almost reach tail tip on perched eagles. Legs are orangish-yellow.

Adults have white heads and tails. Their body and covets are overall dark brown except for *white tail coverts*. Eyes are pale yellow, and beak and cere are orangish-yellow.

Juveniles are overall dark brown except for white axillaries ('wing pits') and a variable amount of white on flight feathers and underwing coverts, usually showing a white median bar. Back and upperwing coverts are tawny-brown and somewhat paler than the dark brown flight feathers, especially noticeable in faded plumage as two-toned uppersides. Tail appears whitish with even-width dark band on tip but can appear all dark on uppersides. Eyes are dark brown, and beak and cere are dark.

Second winter eagles are similar to juveniles but have wide pale superciliaries, whitish bellies that contrast with dark breasts; dark brown upperwing coverts and back show white spots, usually forming a white triangle on latter; and a mix of longer, pointed-tip juvenile and shorter, blunt-tip replacement secondaries form ragged and uneven trailing edges of wings. Eye is medium to light brown, and beak and cere are dark but paler than those of juveniles.

Third winter eagles are similar to second winter eagles except that throat and cheeks are whitish, emphasizing dark ear coverts, and trailing edges of wings are smooth; however some eagles retain one or two longer, pointed juvenile secondaries. Eye is pale yellow, beak is horn-coloured with yellowish areas, and cere orangish-yellow.

Fourth winter eagles are in transition to adult plumage. Head is mostly white but with Osprey-like dark eye-lines, and tips of white tail feathers are usually dusky or black. Brown body and coverts show some white spotting throughout.

Fifth winter eagles are first plumage adults. Many can be recognized by narrow Osprey-like dark eye-lines and dark tips on some tail feathers.

Unusual plumages

Dilute-plumage juveniles and adults have cream-coloured body and covert feathers. Partial albino nestlings and non-adults with some white feathers have been recorded.

Similar species

White-tailed Eagles (Plates 10, 11) appear similar in all plumages. They show seven 'fingers' on wing tips; Bald Eagles show only six. Adults have creamy, not white, heads and necks; dark, not white, undertail coverts; and shorter, more wedge-shaped white tails. Immatures usually do not show paler bellies that contrast with darker breasts and show white spikes on tips of tail feathers.

Flight

Active flight is with slow wingbeats, similar to that of White-tailed Eagle. Soars usually on flat wings but sometimes with a small dihedral. Glides on flat wings with wrists cocked forward. They soar often, many times with other eagles.

Moult

Undergo moult primarily from spring through to autumn, usually suspending it in winter. Replacement of flight and other feathers is not complete annually. Post-juvenile moult begins in March or April of the second calendar year.

Behaviour

Usually found near water where they can find their favourite food, fish. Breeding areas are almost always near water. In winter, when fish may be scarce, they also eat carrion and waterfowl. They regularly pirate food from other raptors, especially other Balds and occasionally Ospreys. The Bald Eagle is a superb fisherman and an agile raptor but prefers to find food in the easiest possible way.

This social species forms communal night roosts, most commonly in winter, of several birds or more and, rarely, several hundred. Individuals often spend considerable time perching. Some birds remain perched in night roosts for a day or two. Balds are vocal, particularly around other eagles, and two eagles often lock talons and whirl with each other (talon-grapple) while in flight.

Status and distribution

Recorded as vagrant twice in Ireland. Fairly common in most areas with large bodies of water in North America north of Mexico. Canadian and Alaskan eagles are migratory, moving to ocean coasts or south into the USA in winter.

Fine points

Immature Bald Eagles differ from immature White-tailed Eagles by contrast between breast and belly colours, whiter wing linings, and narrower wings with only six 'fingers' on wingtip.

Subspecies

The two races, *H. l. alascensis* in Alaska and Canada and *H. l. leucocephalus* in the lower 48 States, differ only by size.

Etymology

In Old English *Balde* meant 'white'; 'Bald-headed Eagle' for 'white headed' and, later, just 'Bald Eagle'. *leucocephalus* is from the Greek *leucos*, 'white', and *kephalus*, 'head'.

Measurements

Length: 70–90 cm (79 cm)
Wingspread: 180–225 cm (203 cm)
Weight: 2.5–6.2 kg (4.5 kg)

Vultures—family Accipitridae

The family Accipitridae consists of all of the kites, eagles, hawks, buzzards, vultures, and harriers.

Vultures are large raptors that are almost exclusively scavengers. On the ground they are ugly and ungainly but in the air they are masters of using air currents for their rather effortless flight.

The five species that occur regularly in the Western Palearctic are all quite different from each other and belong to five different genera. One, the Bearded Vulture, is a denizen of mountains and feeds mainly on bones of large animals. Another, the smallest and most agile, is the Egyptian Vulture. It is more widespread from Spain across the Mediterranean Sea into the Middle East. The third, the Cinereous Vulture, is a large dark vulture with a more restricted range, mainly in Spain and Turkey. The fourth, the Lappet-faced Vulture, is only found locally in Morocco but is more widespread in the Middle East. The last is the Griffon Vulture, a more typical vulture. It is the most common and widespread. A congener, the Rüppell's Vulture, has occurred in the Western Palearctic as a vagrant.

Bearded Vulture   *Gypaetus barbatus*   (Lammergeier)
Plate 12; Photos, p. 302

Description
A large vulture that occurs locally in hilly and mountainous areas of northern Africa, southern Europe, and Turkey. Sexes are nearly alike in plumage and size, but females are somewhat heavier. Four plumages corresponding to age are recognizable. *Flight shape of long narrow pointed wings and wedge-shaped tail is distinctive in all plumages.* Beak is dark horn-coloured.

Adult has white head and underparts that are almost always washed with varying amounts of rufous coloration (usually whiter on heads), as they apply reddish dirt to their plumage. (Adults in zoos almost always appear pure white, probably because they do not have access to reddish soil with which to adorn themselves.) Head also shows wide black masks around each eye, black lores and eye-brows, a narrow black line completely across crown and a short narrow black line on each cheek, and a distinctive tuft of black hair-like feathers that cross the lores and upper mandible and another that extends 4–6 cm below the lower mandible. Eye is yellow, and eye-ring is bright red. A narrow black breast band is usually visible. Upperparts are dark with a distinct greyish cast and narrow pale shaft streaks visible on back, wing coverts, and undersides of flight feathers and tail. Undersides of wings are dark, with coverts appearing darker than flight feathers. Underside of tail is uniformly dark. Adult plumage is acquired at around 5 years of age. The change from juvenile to adult plumage is gradual; however, two intermediate stages, immature (second winter and third winter) and subadult (fourth winter and fifth winter), are recognizable.

Juveniles have blackish heads and necks, with pale brownish-yellow eyes that gradually change to dull yellow or straw-coloured and pale dull brownish-red eye-rings that slowly change to dull red. Their beards are very short, sometimes not noticeable on younger birds, and grow longer with age. Greyish-brown underparts show a variable amount of white spotting and contrast with darker throat and can show a rufous wash. Dark brown backs show extensive whitish mottling, usually forming a triangular patch. Dark brown wing coverts can also show some whitish mottling. Underwings are rather uniformly dark brown; coverts may show some whitish mottling. Secondaries are longer than those of adults and primaries are shorter, resulting in wider shorter wings. Likewise the outer tail feathers are longer than those of adults, resulting in a less deeply wedge-shaped, more rounded tip on tail. The first moult begins when juveniles are about 12 months old.

Second winter birds are similar to juveniles but have a stronger rufous wash on underparts, which also lack white spotting, and begin to show white on their faces. New secondaries are noticeably darker and shorter, with blunter, less pointed tips, resulting in saw-tooth trailing edges of wings. Secondary moult proceeds inward from the outer secondary and outward from the inner. Often a secondary is replaced apparently randomly. Eyes are yellow, and eye-ring is red. Beards are 1–2 cm long.

Third winter birds are somewhat different from juveniles and second winter birds. Their crowns are progressively paler, with darker mask and eye-brows and whitish on nape becoming more apparent. Eyes are yellow, and eye-ring is deep red. Beard is 2–3 cm long. New flight feathers have a greyish cast on their uppersides. Replacement back feathers lack white tips; back becomes progressively darker. Underparts are now a mix of grey, rufous, and white. A few retained juvenile secondaries are visible on the trailing edge of the wings.

Fourth winter birds are much like third winter birds, but with even whiter heads and paler underparts and a thick dark band across the upper breast. Only a few, if any, longer juvenile secondaries are retained. Beard is 3–4 cm long.

Fifth winter birds look almost like adults. Head is almost all white, but still has some dark streaking, heaviest on throat. Eye is yellow and eye-ring is deep red. Beard is over 4 cm in length. Back and wing coverts are now black; new feathers have a greyish cast and pale shaft streaks, and retained ones are whitish-brown. Most flight and tail feathers have greyish cast to their uppersides. Underparts are now white with a rufous cast. All juvenile secondaries have been replaced, so that trailing edges of wings are now smooth. Tail shape as in adult. Uppertail coverts show some brownish feathers.

Unusual plumages

No unusual plumages have been reported. However, birds released in the Alps show individually distinct patterns of bleached flight feathers.

Similar species

Egyptian Vulture (Plate 13) juveniles and first summer birds can also appear overall dark with a similar wedge-shaped tail like juvenile and first summer Bearded Vultures but are much smaller and have less pointed wingtips and relatively shorter wings and tails.

Flight

Wonderful fliers, using updrafts and slope life for their effortless gliding on slightly drooped wings while foraging, usually gliding rather close to the mountainous terrain. They soar occasionally, with wings held level. Powered flight is with slow heavy wing beats, and usually only used to take off. They also hover occasionally, just before dropping bones and when looking over possible food.

Moult

Adults undergo an annual moult after breeding, usually beginning in April. Most feathers are replaced every 2 years. Primaries are replaced with a wave moult. Juveniles begin moult at around 12 months of age. The moult of older immatures is similar to that of adults.

Behaviour

Scavengers that spend much time regularly foraging over their hilly or mountainous territory searching for fresh carrion. Most of their diet is bones, but they also consume meat, preferring freshly dead animals. Bones are eaten whole, except for the largest bones, which are taken aloft in their feet and dropped on a large rock to break them apart, after which the fragments are consumed.

They usually occur singly or in pairs, unlike the gregarious Griffon and Egyptian Vultures; however, small groups of young immatures are sometimes encountered.

Large stick nests are built in a small cave or on a ledge with a rock overhang. Courtship includes undulating flights and mutual preening.

They are usually silent, but calls reported include whistling during display flights.

Status and distribution

Uncommon local permanent residents of hilly and mountainous areas of the Western Palearctic, specifically of the Atlas Mountains of Morocco and barely into north-western Algeria, and of the Pyrenees Mountains of Spain and France; now being reintroduced into the Alps of southern Europe and into northern Greece, and widespread across Turkey into the Caucasus. They bred formerly in Israel and may still be present locally in Sinai and Egypt.

The reintroduction program in the Alps has been successful in putting out many young birds. A few of these have wandered far afield, with sightings of two in Holland.

Fine points

All Bearded Vultures regularly douse themselves with iron oxide laden dirt on body feathers, which gives them their rufous cast. Amount of rufous varies with individuals and most probably reflects the availability of this type of soil, rather than geographic variation, as suggested by some authors.

Subspecies

Hiraldo *et al.* (1984) make a good argument for considering that all Western Palearctic Bearded Vultures are of the same race, *G. b. barbatus*, incorporating *G. b. aureus*. They conclude that the only other recognizably different race is *G. b. meridionalis* of East and Southern Africa.

Bearded Vulture

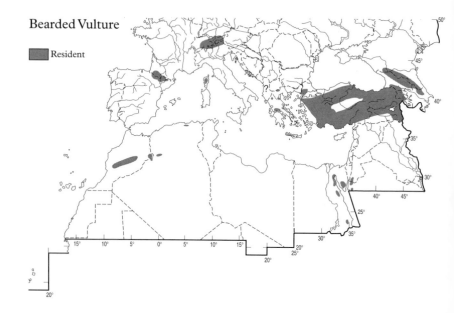

Etymology

Gypaetus comes from the Greek *Gyps* for 'vulture' and *aetos* for 'eagle'. *barbatus* is Latin for 'bearded'. *Lammergeier* is the former German common name and meant 'Lamb vulture'. This was changed to *Bartgeier* (Bearded Vulture) recently for conservation reasons.

Measurements

Length: 1.0–1.2 m (1.1 m)
Wingspread: 2.4–2.9 m (2.7 m)
Weight: 4.2–6.5 kg (5.4 kg)

Egyptian Vulture   *Neophron percnopterus*

Plate 13; Photos, p. 300

Description

The Western Palaearctic's smallest vulture occurs primarily around the Mediterranean basin. Sexes are nearly alike in plumage and size. Five plumages corresponding to age are recognizable. *Long narrow beak and wedge-shaped tail are distinctive in all plumages.* A small patch of yellow (adult) to grey (juvenile) bare skin is visible in centre of breast when crop is full. Irises are dark brown. Beak is dark. Leg is dull pinkish.

Adult. Bare face skin, base of lower mandible, and elongated cere are, according to one authority, yellow on females and orange-yellow on males (at least when breeding). Crown is covered by short white down. *Nape and neck have long, lanceolate white feathers.* A black smudgy line often occurs under each eye (according to one authority, only on males), but, as it is present on many females and absent on many males, it is not a sex character. Body, tail, and underwing, secondary upperwing, and tail coverts are completely white, but underparts, head and neck, back, and some secondary upperwing coverts and scapulars are usually sullied brown, sooty, or rust; this apparently done on purpose. (Adults in zoos almost always appear pure white, probably because they do not have access to mud, iron-ore-rich soil, or burned material with which to adorn themselves.) Primary upperwing coverts and primaries are black. Outer webs of secondaries are white except for dark bases and tips; this results in two narrow dark bands across the whitish uppersides of secondaries. Whitish outer webs of inner primaries are visible as rays on upperwings. Inner three greater secondary upperwing coverts on each wing are usually dark. Flight feathers are black on undersides and contrast with white underwing coverts.

The four immature annual plumages are described below in chronological order.

Juvenile. Head consists of grey bare face skin, pale yellow basal half of dark beak, dark down on crown that comes to a point in a distinctive 'widow's peak' on fore crown, and blackish nape and neck feathers that are shorter than those of adults. Body and wing coverts vary from grey-brown to brown to brownish-black, with a variable amount of pale diamond-shaped markings. Colorations of upperparts and underparts are quite variable; some birds appear overall creamy on underparts or upperparts or both. Undertail and uppertail coverts are creamy and contrast with darker body. Bare crop patch on chest is pale grey. Flight feathers are dark brown, with a pale brown cast to uppersides of secondaries. Tail is greyish brown with paler tip. Legs are pale grey.

Second winter birds are essentially like juveniles, except that new uppertail and undertail coverts are the same brownish colour as body feathers, uppersides of new flight feathers have a greyish cast, upperparts lack pale diamond-shaped markings, and forehead is covered with a mixture of new short whitish down and old blackish down, lacking dark 'widow's peak'. Bare crop patch on chest of older immature is most often the same colour as their faces.

Plumages of third and fourth winter vultures are quite variable.

Third winter birds are similar to younger birds but have a variable number of new white feathers on median underwing coverts and axillaries and new whitish body and covert feathers, giving overall a mottled or piebald appearance, with a noticeable white bar across each underwing. Greater underwing coverts remain completely dark. Whitish crown and dark nape feathers give head a monkish appearance. Face skin is usually pale yellow. Secondaries and inner primaries have greyish cast on outer webs, resulting in a pale band across uppersides of flight feathers. New tail feathers are frosty-grey.

Fourth winter birds are somewhat similar to third winter ones but are overall more whitish and adult-like, including having a few new white or partially white tail feathers. Greater underwing coverts are usually mostly white. Neck is usually still quite dark, forming a dark collar. Small black patches next to body on upperwings (inner upperwing greater secondary coverts) may or may not be present in this plumage. Face is yellow.

Fifth winter birds are essentially like adults except for progressively fewer retained darker feathers on body, neck, coverts, and tail. Face skin, base of beak, and legs are usually yellow.

Unusual plumages

No unusual plumages have been reported.

Similar species

Bearded Vulture (Plate 12) juveniles and first summer birds are also dark with wedge-shaped tail but are much larger with a relatively longer tail and longer, more pointed wings.

Booted Eagle (Plate 37) light morph has pattern similar to that of adult Egyptian Vulture (black flight feathers and white body and coverts) but has square-tipped, not wedge-shaped tail tip and has dark uppersides.

Flight

Soar with wings level or slightly cupped and glide with wings slightly cupped and tips held lower than body. Active flight is with strong, deep wing beats of stiff wings.

Moult

Annual moult is usually not complete, with some primaries and secondaries retained. Subsequent moult of flight feathers begins anew as well as continuing where old one left off. Post-juvenile moult begins in spring with body moult, and continues into the summer and early autumn. Not all feathers are replaced each year. Subsequent moults are like those of adults. Adult plumage is attained in 4 to 5 years.

Behaviour

Primarily scavengers but utilize a wide variety of food sources, including capture of live prey such as turtles and insects. They are less dependent on large carcasses than are large vultures and feed on these after other vultures have eaten the soft parts. A variety of small mammals, birds, reptiles, and insects are reported as staple food items. They also feed on birds' eggs. Using stones to break large eggs (e.g. those of Ostrich) is an interesting use of tool; they break smaller eggs by lifting them in their beak and smashing them on the ground.

Stick nests are built on ledges and large potholes in cliffs and are lined with a variety of materials, many of which are man-made; however, nests on buildings and in trees have been reported. Nests are often in close proximity. Courtship includes undulating flights, and mutual preening of pairs occurs often. They are usually silent, but mewing and shrill trilling calls have been reported, and groans and grunts have been heard around carcasses. Young are fed whole, not regurgitated, food.

Communal night roosts are formed, usually on cliffs.

Status and distribution

Egyptian Vultures are summer residents of open, often arid areas around the Mediterranean basin, with population centres in Spain,

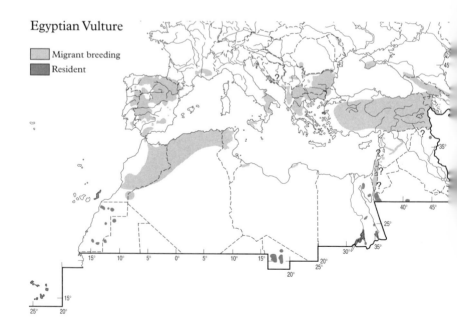

Egyptian Vulture

Migrant breeding
Resident

north-western Africa, Turkey, and parts of the Middle East and smaller, local populations in Portugal, France, Italy, the former Yugoslavia, Greece, Cyprus, Romania, Bulgaria, Georgia, and the Caucasus. They concentrate at food sources such as rubbish dumps, slaughterhouses, and fishing ports.

Most birds migrate, wintering south of the Sahara, but a few individuals are reported in Europe almost every winter. In late spring or summer, some wander afield into northern Europe. A few immatures remain on winter grounds during the northern summer. Small populations in the Canary and Cape Verde Islands are resident.

Fine points
Egyptian Vultures in Pakistan, Nepal, and India (*P. n. ginginianus*) have yellow beaks.

Subspecies
Neophron p. percnopterus occurs in the Western Palearctic.

Etymology
Neophron was a character from the pseudo-Greek mythological *Metamorphoses*; he was changed into a vulture by Zeus; *percnopterus* comes from the Greek *percnos* and *pteros* for 'dusky' and 'wing'.

Measurements
Length: 55–65 cm (60 cm)
Wingspread: 148–171 cm (159 cm)
Weight: 1.4–2.3 kg (1.8 kg)

Rüppell's Vulture   *Gyps rueppelli*

(Rüppell's Griffon Vulture/Rüppell's Griffon)

Plate 14

Description

A large dark African vulture that has occurred as a vagrant in Egypt. In flight, *Gyps* vultures have a distinctive shape: short head, wide wings showing a secondary bulge, and short square tail. In all plumages Rüppell's are distinguished from the similar Griffon Vulture by dark body and wing and tail coverts and by wings held more level when soaring. They are, like the Griffon, social, gathering at carcasses in numbers. Adults, immatures, and juveniles differ sufficiently to be distinguishable in the field. Sexes are alike in plumage and size.

Adult head and neck are dark and sparsely covered with short with down, heavier on the crown and nape. Lores and cere are blackish, eyes are yellow, and beak is orangish-yellow, with a black tip. Body, wing and tail coverts, and outer leg feathers are dark brown, with distinctive wide creamy white tips giving an overall scaly appearance; tips are wider on underparts and upperwing coverts. Some adults (older?) have uniform whitish to creamy underparts. Inner legs are covered with a whitish down, obvious on perched and landing adults. Flight and tail feathers are dark brown. *Gyps* vultures have a ruff on the upper back at the base of the neck. On adults this is composed of short white feathers in a tightly packed cottony patch. The crop, when full, is visible as a dark bulge of skin on the breast below the base of the foreneck. *Gyps* vultures also have two distinctive circular bare patches of skin on either side of the crop, which vary in colour from pale bluish to dark red. These are thought to be used for signalling and are surrounded by a circle of tightly packed cottony white down. Dark undersides of wings show a narrow whitish bar in the lesser underwing coverts and several incomplete narrow white lines formed by the pale tips of the median and greater coverts. Legs and feet are dull grey.

Juveniles are somewhat different from adults in plumage. They are overall dark brown, lacking any pale feather edging; however, underparts show pale streaking, and upperwing coverts and back feathers have dark tawny central streaks. Their beaks and ceres are dark, and their ruffs are composed of long, lancelot, dark brown feathers. In flight, the pointed tips of the dark secondaries form a serrated trailing edge to the wings. Underwings are dark except for the narrow whitish bar in the lesser underwing covert region. Eyes are dark brown.

Older immatures gradually change from juvenile to adult plumages. The first noticeable changes are the replacement of pointed-tipped with blunt-tipped secondaries. New upperwing coverts have buffy tips and lack the central streak. Bill begins to lighten, beginning with the culmen and gradually expanding. Eyes begin to lighten. The replacement ruff feathers become shorter with whitish centres after a few years but are still lanceolate. Adult plumage is attained after 5 or 6 years; however, adults may continue to become paler with age, especially obvious on underparts and upperwing coverts.

Unusual plumages
No unusual plumages have been reported.

Similar species
Griffon Vulture (Plate 14) is similar in size and shape to the Rüppell's Vulture but has more uniform, yellow-brown body and coverts that contrast with darker flight feathers in all plumages. They soar with wings held in a dihedral; Rüppell's soar with wings more level.

Cinereous Vulture (Plate 15) is also a large, uniformly dark vulture but is larger, more eagle-like. It lacks the S-shaped curve to the trailing edge of wings and white patagial bar on underwings.

Flight
Active flight is with slow, deep, powerful wingbeats. Soars and glides with wings held level. Long neck is folded and tucked into the body

during flight, resulting in a short head projection. *Gyps* vultures spend much of the day on the wing, sometimes at great altitude, whenever thermals, slope lift, or winds sufficient for dynamic soaring are available.

Moult
Like most African raptors, moult continues year round. It takes more than 1 year to complete. Moult may be suspended while breeding by adults. Post-juvenile moult begins at about 10 months of age.

Behaviour
Scavengers that spend much time in flight searching for food, often at great altitudes. They gather at medium to large carcasses, sometimes in large numbers. They subsist almost entirely on carrion.

They nest in colonies of a few to hundreds of pairs on large cliffs, constructing large stick nests on ledges. One chick is raised annually. Courtship consists mainly of mutual flying of pair in front of and above nesting cliffs.

Vocalizations of grunts, hisses, and chattering are often heard at carcasses during dominance interactions. Necks of aggressive vultures often become deep red.

Status and distribution
Have been recorded as vagrants at least three times in Egypt and four times in Spain.

Fine points
While gliding, *Gyps* Vultures will sometimes bend their wing downward until the wingtips almost touch, in an action called a 'flex'.

Subspecies
Gyps r. erlangeri of Ethiopia and Somalia should be the race recorded in Egypt, but the two races differ little, if any, in plumage.

Etymology

Gyps comes from the Greek for 'vulture'; *rueppelli* named for Dr Eduard Rüppell. Also called Rüppell's Griffon.

Measurements

Length: 90–113 cm (101 cm)
Wingspread: 225–250 cm (236 cm)
Weight: 5.5–9.1 kg (7.6 kg)

Griffon Vulture   *Gyps fulvus*

(Eurasian Griffon)

Plate 14; Photos, p. 304

Description

A large pale vulture that occurs in southern Europe, north Africa, and the Middle East. In flight, *Gyps* vultures have a distinctive shape: short head, wide wings showing a secondary bulge, and short square tail. Griffons soar with wings held in a dihedral. In all plumages, body and pale wing and tail coverts contrast with darker flight and tail feathers. They are decidedly social, gathering at carcasses in numbers and breeding in colonies. Adults, older immatures, and juveniles differ sufficiently to be distinguishable in the field. Sexes are alike in plumage and size.

Adult head and neck are covered with short white down, except for two rarely seen small bare areas at the base and on either side of the hind-neck. Lores and cere are blackish, eyes are yellow-brown, and beak is yellow (with black on cutting edges). Body, wing and tail coverts, and outer leg feathers are pale to creamy yellow-brown; underparts show pale white streaking. Inner leg feathers are whitish. Flight and tail feathers are dark brown. *Gyps* vultures have a ruff on the upper back at the base of the neck. On adults this is composed of short white feathers in a tightly packed cottony patch. The crop, when full, is visible as a dark bulge of skin on the breast below the base of the foreneck. *Gyps* vultures also have two distinctive circular bare patches of skin on either side of the crop, varying in colour from pale bluish to dark red. These are thought to be used for signalling and are surrounded by a circle of tightly packed cottony white down. Undersides of wings show two pale bars on the wing linings: a narrow white one in the patagium, and the other a more or less defined wide band across the greater secondary coverts; occasionally other areas are pale as well. Undersides of pale secondaries have narrow dark tips and dusky outer webs, which appear as faint streaks. Greater secondary upperwing coverts usually have dark brown centres. Legs and feet are dull grey.

Juveniles are similar to adults, but they are overall a somewhat darker brown, their beaks and eyes are dark, and their ruffs are composed of long, lanceolate, tawny-brown feathers. Their underparts, upperwing coverts, and outer leg feathers show pale central streaks, obvious on perched vultures. In flight, the pointed tips of the dark secondaries form a serrated trailing edge to the wings.

Older immatures gradually change from juvenile to adult plumages. The first noticeable changes are the replacement of pointed-tipped with blunt-tipped secondaries; trailing edge of wings now appear uneven and ragged. New upperwing coverts are paler and lack the white central streak; coverts appear overall more mottled. Bill begins to lighten, beginning with the culmen and gradually expanding. Eyes begin to lighten. The replacement ruff feathers become shorter and whitish after a few years but are still lancelot. Adult plumage is attained after 5 or 6 years; however, adults may continue to become paler with age.

Unusual plumages

No unusual plumages have been reported. Reintroduced vultures in Italy have individually distinct patterns of bleached flight feathers.

Similar species

Cinereous Vulture (Plate 15) is somewhat similar; see under that species for distinctions.

Flight

Active flight is with slow, deep, powerful wingbeats. Soars with wings held above the horizontal in a dihedral; glides with wings held level, sometimes with wrists raised a bit. Long neck is folded and tucked into the body during flight, resulting in a short head projection. Griffon Vultures spend much of the day on the wing, sometimes at great altitude, whenever thermals, slope lift, or winds sufficient for dynamic soaring are available.

Moult

Annual moult is not complete and is suspended during winter and, for adults, while breeding. More than half of the body and coverts feathers are replaced annually, whereas less than half of the flight and tail feathers are. Post-juvenile moult begins during the spring of second calendar year.

Behaviour

Scavengers that spend much time in flight searching for food, often at great altitudes. They gather at medium to large carcasses, sometimes in large numbers. They subsist almost entirely on carrion.

They nest in colonies of a few to hundreds of pairs on large cliffs, constructing large stick nests on ledges. One chick is raised annually. Courtship consists mainly of mutual flying of pair in front of and above nesting cliffs.

Vocalizations consist of grunts, hisses, and chattering and are usually heard only at carcasses during dominance interactions. Two bare spots on breast of aggressive vultures become dark red.

Status and distribution

Permanently resident near and around hills and mountains in many areas: locally in Atlas Mountains of Morocco and Algeria; most of the Iberian Peninsula; Massif Central of France; Sardinia; locally in the former Yugoslavia and northern Greece; Cyprus; southern Turkey; and locally in the Caucasus, north Israel and Jordan, and southern Syria. Breeding vultures in most of Turkey are migratory, leaving in winter. Non-breeding vultures sometimes move far afield from nesting colonies. Juveniles have shown up as vagrants in most countries of central Europe, as well as in southern Sweden.

Populations in most areas are declining, thought to be due to mass poisoning, electrocution, direct persecution, and altered husbandry practices. Populations in France and Italy are being restored by reintroduction.

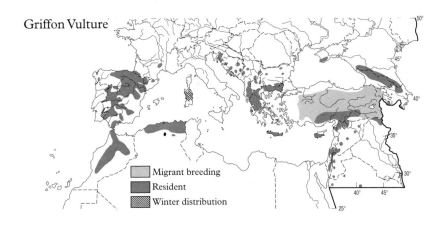

Griffon Vulture

Migrant breeding
Resident
Winter distribution

Fine points

While gliding, Griffon Vultures will sometimes bend their wing downward until the wingtips almost touch, in an action called a 'flex'.

Subspecies

Gyps f. fulvus occurs in the Western Palearctic.

Etymology

'Griffon' comes from the Greek *gryphos* for 'hook-nosed', originally the name of a legendary bird that was half eagle and half lion, but later in the 17th century, used in French to mean vulture. *Gyps* comes from the Greek for 'vulture'; *fulvus* is Latin for the colour tawny or yellow-brown.

Measurements

Length: 93–117 cm (105 cm)
Wingspread: 232–276 cm (254 cm)
Weight: 5.1–8.5 kg (6.8 kg)

Cinereous Vulture   *Aegypius monachus*

(Eurasian Black Vulture/Black Vulture)

Plate 15; Photos, p. 306

Description

This vulture of southern Europe and the Middle East is the Western Palearctic's largest bird of prey. Flying birds lack the secondary bulge of Griffon Vultures; front and rear edges of wings are nearly parallel. Trailing edge of wing appears serrated because secondary tips are pointed. When gathered at carcasses with Griffon Vultures, their larger, square heads are distinctive. Tails of flying birds are wedge-shaped. Pale legs and feet are noticeable on close birds. Adults and juveniles are similar in plumage; juveniles are overall darker. Females average larger than males.

Adult is dark brown, except for head and neck, which are mostly covered with pale down but with strongly contrasting *black throats and black areas around each eye*. Buffy breast shows some dark brown streaking. Hind-neck has an 'Elizabethan' brown ruff. Up close, pale ruff feathers are noticeable, as are paler backs and upperwing coverts on some birds. In flight the underwing shows flight feathers somewhat paler than coverts, with even paler bases that form a narrow pale line through underwings. Cere and bare areas around mouth are pale blue-grey and pink; *dark beak shows a small horn-coloured area next to cere*. Iris varies from hazel, golden-brown, to reddish-brown.

Juveniles are darker than adults, appearing almost black. Head is covered almost completely with short black down. Underwings lack the narrow pale line seen on adults. Cere and bare face skin vary from pinkish to pale grey; beak is solid dark. Iris changes progressively from dark brown to reddish-brown. As birds age, they gradually become paler, progressively showing more pale down on the head, but all have black throats and areas around eyes.

Unusual plumages

Birds with some white feathers on greater secondary upperwing coverts are seen regularly. According to one authority, this may be a character of males.

Similar species

Lappet-faced Vulture (Plate 15) of Morocco, Israel, and Jordan is also a large dark vulture. For differences see under 'Similar species' in that account.

Griffon Vultures (Plate 14) are slightly smaller, have pale body and wing coverts that contrast with flight and tail feathers, and show a strong 'S-shaped curve' on trailing edge of wing. Wings are held above horizontal while soaring and level when gliding.

White-tailed Eagle (Plate 11) juveniles can also appear uniformly dark in flight but have longer head projection and usually show white on axillaries and undertail.

Greater Spotted Eagle (Plates 29, 30) can appear similar in flight but are much smaller and have longer head and tail projections and relatively shorter wingspan.

Flight

Powered flight is with slow, deep, powerful wingbeats. Soars with wings flat; glides with inner wings level and wingtips held below wrists.

Moult

Annual moult is not complete and is suspended during winter and while breeding. Post-juvenile moult not recorded.

Behaviour

Scavengers that spend much time in flight searching for food. They gather at medium to large carcasses with other vultures, over which they dominate. They are less social than other species and usually only one will feed at a time. They have a curious behaviour of bending over with body horizontal and erecting special breast feathers; these stand up like two brushes on each side of head. This is thought

to be an aggressive signal to other vultures and is often used with slow foot showing display, similar to that of Griffon Vultures (see also 'Fine points').

Cinereous Vultures are not above pirating prey from other raptors.

Status and distribution

Resident in west central Spain, northern Majorca, and from northern Greece across Turkey into the Caucasus. Scattered pairs are recorded in Bulgaria, Cyprus, and elsewhere. Immatures have shown up as vagrants in many areas of the Western Palearctic.

Fine points

Cinereous Vultures appear distinctive as they descend to carcasses, with wings drooped, 'brushes' erected, and tail cocked up sharply. Final approach is characterized by swept back wings, head low, and feet thrown forward level with head.

Subspecies

None.

Etymology

Cinereous comes from the Latin for 'ashy'. *Aegypius* is Greek for 'a vulture'. *monachus* is the Latin for 'hooded', i.e. like a monk, and

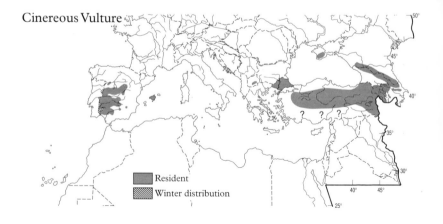

Cinereous Vulture

Resident

Winter distribution

was derived from the Greek *monakhos* for 'a monk', from *monos*, 'alone'.

Measurements

Length: 100–115 cm (108 cm)
Wingspread: 265–290 cm (280 cm)
Weight: 6–13 kg (10 kg)

Lappet-faced Vulture   *Torgos tracheliotus*

Plates 15, 47; Photos, p. 302

Description

Two races occur in the Western Palearctic: *negevensis* is a local resident of arid acacia savannahs of the Middle East and *tracheliotus* is a rare and local resident of Morocco. They are similar in size and plumage to Cinereous Vulture, but *undertail coverts are completely pale and undersides and underwings show whitish areas.* Flight feathers are a bit paler than coverts, and markings on secondaries form streaks on underwing. Trailing edge of wings appears serrated because of pointed secondary tips. Perched birds show huge beak and squarish head. Lappets on side of neck are small and indistinct in *negevensis*, unlike those of African birds. Adults and juveniles are similar in plumage. Sexes are alike in plumage; but females average somewhat larger and quite a bit heavier. Iris is dark brown. *Beak is two-toned: dark on top with variably sized pale area below.* Cere, feet, and legs are blue-grey.

Adult head and front neck are pale pink (*negevensis*) to red (*tracheliotus*); hind-neck has an 'Elizabethan' brown ruff. Back and upperwing coverts are brown and contrast somewhat with uppersides of dark brown flight feathers. According to one authority, males often show some white feathers on back and upperwing coverts. Underparts and underwing coverts are brown with some whitish mottling. Flight feathers below are a bit paler than coverts, with pale bases that form a pale line through underwing. The races differ in the size of the lappets, larger in *tracheliotus*; in the face colour of adults, red in *tracheliotus* and pink in *negevensis*; and in the thigh colour of adults, brown on *negevensis* and white on *tracheliotus*. Tail is dark brown.

Juvenile is similar to adult but head is covered with whitish down, bare neck skin is livid white to pale blue-grey, and plumage is darker overall, with less whitish mottling on underparts.

Unusual plumages

White speckling on the back also occurs (see above), according to another authority, on immatures. A partial-albino adult from Kenya had mostly white upperwing coverts and back.

Similar species

Cinereous Vulture (Plate 15) is also a large dark vulture with whitish head but has more uniformly coloured upperparts, black mask on head, longer tail, and dark undertail coverts; droops wingtips when gliding; and lacks 'S-shaped curve' on trailing edge of wing when gliding. On perched birds, pale area on beak is smaller and more localized.

Griffon Vulture (Plate 14) is smaller with pale body and wing coverts and soars with wings in dihedral.

White-tailed Eagle (Plate 11) juveniles can also appear dark with whitish areas but are smaller, have longer head projection, and usually show whitish axillaries and undertail.

Flight

Powered flight with slow deep powerful wingbeats. Soars and glides with wings level.

Moult

Annual moult is not complete, as is to be expected in large raptors. Apparently in moult throughout year, most likely suspended while breeding.

Behaviour

Scavengers that thus spend much time in flight searching for food. But they are also often seen perched atop acacia trees. They gather at carcasses with other vultures, over which they dominate. Usually they assert themselves at carcass with 'threat-walk'. They have been reported to kill live prey.

Lappet-faced Vulture

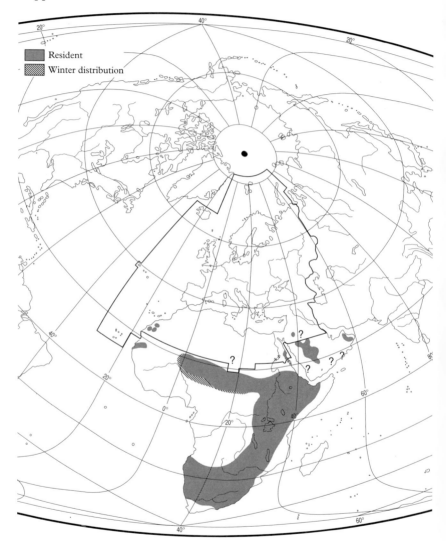

Status and distribution

Rare and local resident in arid acacia savannahs of southern Israel (where now extirpated), and most likely in adjacent areas of Sinai, and Jordan. Main population of *negevensis* is throughout the Arabian peninsula. Race *tracheliotus* is a rare and local resident in central and southern Morocco. Formerly much more widespread in north-west Africa, occurring in Tunisia, Algeria, and Mauritania.

Fine points

Lappet-faced Vulture's large beak widens in front of cere; that of Cinereous Vulture gradually narrows in front of cere.

Subspecies

T. t. negevensis occurs in the Middle East. *T. t. tracheliotus* occurs north-west Africa. Some authorities place this species in the genus *Aegypius*.

Etymology

Torgos is Greek for 'vulture'; *trachelos* is Greek for 'throat', apparently for its naked throat, and *otos* is Greek for 'ear', in reference to its lappets.

Measurements

(race *tracheliotus* somewhat smaller)
Length: 96–115 cm (108 cm)
Wingspread: 258–295 cm (280 cm)
Weight: 5–14 kg (10 kg)

SNAKE EAGLES—GENERA *CIRCAETUS* AND *TERATHOPIUS*

Only one snake eagle, the Short-toed, occurs as a summer breeder in the Western Palearctic. Snake eagles are large pale eagles that specialize in preying on snakes and other reptiles, often hunting by hovering. Their heavily scaled legs protect them from bites from their prey.

A closely related snake eagle, the Bateleur, occurs in the Middle East as a rare vagrant from sub-Saharan Africa.

Circaetus is from the Greek *Kirkos* for 'harrier' and *aetos* for 'eagle'. The other five species in the genus are resident in parts of sub-Saharan Africa.

Short-toed Snake Eagle   *Circaetus gallicus*

(Short-toed Eagle)

Plate 5; Photos, p. 308

Description

A distinctive pale-bodied medium-sized snake eagle, widespread through the southern parts of the Western Palaearctic. Large head gives perched birds an owl-like appearance. In flight its *pale underparts, square-tipped tail, and dark head and upper breast are diagnostic*. Wings are proportionally large and, when spread, are widest at the wrist. Plumages of the sexes are almost alike; females on average are larger than males. Adults and juveniles are similar in plumage. On perched birds, wingtips reach tail tip. Legs are dusky-yellow.

Adult's head is usually dark brown, often with a greyish cast, but head can be paler and is completely pale on some birds. Upperparts are uniformly dark brown, except for median upperwing coverts, which are paler brown and form a wide pale band across each upperwing. Breast is dark brown, forming a bib; males usually have vertical pale streaks in the breast, females usually do not. However, pale-headed eagles usually only show a dark incomplete Osprey-like necklace on upper breast in place of dark bib. Belly, leg feathers, and undertail coverts are white; eagles with dark heads usually have dark barring on belly. Underwing is whitish and varies from a strong pattern of five or six dark bands on darker birds to being almost unmarked on paler birds. Upperside of brown tail appears dark but underside is paler with three dark brown bands. Iris is lemon-yellow to orange-yellow.

Juvenile differs from adult by rufous-brown, rather than dark brown, breast coloration, by less banding on the undersides of secondaries, especially lacking the outermost dark band, and by a wider pale terminal tail band. Iris is lemon-yellow.

Unusual plumages

No unusual plumages have been reported.

Similar species

Ospreys (Plate 5) may appear similar in flight; see under that species for distinctions.

Bonelli's Eagle (Plate 36) juveniles in spring can fade to creamy-whitish on undersides and appear similar to large, pale, dark-headed eagles, but their secondaries are darker than primaries on underwing, and they usually show a narrow dark line on each wing at base of secondaries (dark tips of greater secondary underwing coverts).

Honey Buzzard (Plate 6) pale adults also show strong bands on underwing, but are smaller, have dark carpal patches on underwing, and lack dark hooded look.

Honey Buzzard (Plate 6) pale juveniles appear similar but are smaller and have dark carpal patches and secondaries darker than primaries on underwings.

Flight

Powered flight is with deep, slow, powerful wingbeats on flexible wings. Soars with wings flat or in a slight to moderate dihedral; glides with wrists up and pushed forward and wingtips pointed down and back and often with a modified dihedral, appearing somewhat harrier-like. Hovers and kites for considerable periods of time searching for prey, sometimes at great altitudes and often with feet dangling.

Moult

Not completely understood, but most probably eagles replace most or all body feathers, coverts, primaries, and rectrices every year, but only some of the secondaries. Moult takes place both during summer on breeding grounds and in winter in Africa, but is suspended during migration. Spring juveniles in Israel were replacing inner primaries in early May.

Behaviour

Prey primarily on snakes and lizards, but some mice and nestling birds are also taken. Usual hunting technique is aerial, with much time spent hovering or kiting in one place or gliding slowly over the ground, often at great altitudes. They will also hunt from high perches, including man-made ones, particularly power poles. Individuals may sit motionless for long periods on a perch or the ground.

Threat flight at intruding conspecifics is with head and long neck extended upwards, wings in a shallow upturned 'U', legs dangling, and tail cocked somewhat upward; this is accompanied by loud calling. This species is vocal usually only during the breeding season. Typical raptor nests are built in rather low trees but are rather small for such a large bird. Only one young is raised annually.

Status and distribution

Summer residents, breeding in open areas of northern Morocco and Algeria, Spain, southern France, locally in Italy and the former Yugoslavia, Greece, Turkey, Lebanon(?), Israel, southern Caucasus, Ukraine, Russia, Belarus, and eastern Poland. Range formerly extended into northern France, Belgium, Germany, Denmark, and Switzerland. Most of the population migrates to Africa south of the Sahara in a band from Senegal to Ethiopia to spend the northern winter; however, there are winter reports from Sicily (where over five are recorded annually) and elsewhere in southern Europe. Large numbers are recorded on migration at Gibraltar, Bosphorus, and Israel in September and October; spring counts in April and May are somewhat smaller. Birds from Asian breeding range winter on the Indian subcontinent.

Fine points

Differences in breast pattern by sex have been reported—solid dark breast on females and whitish on breasts of males—but some exceptions occur.

Subspecies

None. Thought to form a superspecies with *C. pectoralis* and *C. beaudouini* of sub-Saharan Africa.

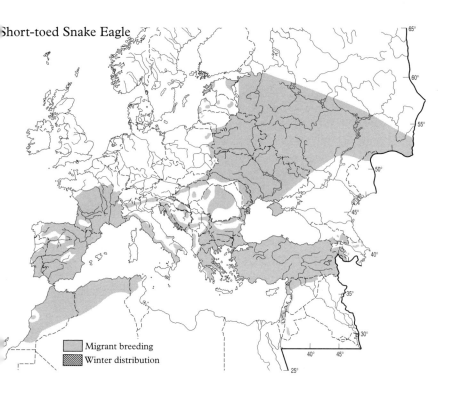

Short-toed Snake Eagle

Migrant breeding
Winter distribution

Etymology

gallicus is Latin meaning 'from France'.

Measurements

Length: 59–65 cm (62 cm)
Wingspread: 165–188 cm (177 cm)
Weight: 1.2–2.3 kg (1.7 kg)

Bateleur   *Terathopius ecaudatus*

Plate 48

Description

A distinctive Snake Eagle, a rare vagrant to the Middle East from east Africa or southern Arabia. Its flight silhouette of long narrow wings held in a dihedral and short tail is unlike that of any other raptor, rather like a 'flying wing'. Five age-related plumages can be recognized in the field. Adult plumage is most probably reached in 5 years. Sexes are alike in plumage, except for adults and older immatures. Wing tips extend beyond tail tip on perched eagles, when elongated nape feathers are often raised into a short, cowl-like wide crest. Eye colour is brown to dark brown. Bare face skin colour varies with age.

Adults have black heads with bright red face skin and yellow beak with dark tip. Body is overall black except for chestnut or, rarely, creamy-white back and pale grey upperwing coverts; these are separated by black scapulars. Underwing coverts are white. Males have all-black secondaries and inner primaries; these are white with narrow black tips on females. Outer primaries of both are whitish below with dark tips but are dark above. Leg feathers are black. Tail coverts and extremely short tail are chestnut. Legs and toes are bright red and extend well beyond short tail on flying eagles.

Juveniles are overall medium brown, with noticeably paler head, especially throat. Buffy feather edges are noticeable on underparts in fresh plumage. Face skin is greenish-yellow. Base of beak is yellowish and tip is dark. Tips of greater underwing coverts are buffy and form a narrow pale band across each underwing. Undersides of secondaries appear uniformly brown. Legs are whitish to pale yellow. Brown tail is much longer than that of adults; legs fall somewhat short of tail tip on flying eagles.

Second plumage eagles or **immatures** are almost identical to juveniles, as colours of soft parts have not changed and new replacement feathers are the same colour or at most slight darker. They differ in having shorter tails, such that the toes reach the tail tip on flying eagles; in having shorter secondaries, thus narrower wings; and in showing signs of moult when seen close. New secondaries below appear pale with dark tips.

Third plumage eagles or **subadults** are overall blackish-brown, including back and upperwing coverts, with pinkish face skin and adult-like beak. Underwings show dark coverts and pale flight feathers with wide dark tips. Chestnut tail is much shorter so that toes protrude well beyond tail tip on flying eagles. Legs are yellow to orange.

Eagles in transition to adult plumage show whitish mottling on underwing coverts and underparts mottled black and blackish-brown. Characteristic adult male or female undersides of secondaries begin to appear, as do chestnut back and pale grey upperwing coverts. Face skin and legs become red.

Unusual plumages
No unusual plumages have been recorded.

Similar species
No raptors in the Western Palearctic are similar.

Flight
Bateleurs' long narrow wings and short tails are adapted for efficient high-speed flight. Powered flight is with rapid, shallow wingbeats and is used mainly for take-offs. They soar and glide with wings in a strong dihedral, usually rocking from side to side.

Moult
Not studied well. Apparently, like most tropical raptors, they undergo a slow moult throughout the year. Juveniles begin moulting into the next plumage at approximately 1 year of age. Adult plumage is most probably attained after 5 years.

Behaviour
Specialized aerial raptors that spend most of the day on the wing hunting. They take a wide variety of birds and mammals, some rather

large, and smaller numbers of snakes and lizards, and even a few fish. Most prey are captured after often tremendous stoops, either on the ground or in the air (birds). Carrion is often eaten, especially by juveniles and older immatures, and prey is regularly pirated from other raptors.

Status and distribution
A rare vagrant to the Middle East from Africa or southern Arabia, with one record from Iraq (October) and six from Israel (spring).

Fine points
The bare face skin is thought to be used by adults for communication of mood, as it varies from yellow to pink to bright red.

Subspecies
Monotypic.

Etymology
Bateleur is a French word, no longer in use, for circus-like performers who travelled the countryside juggling, tumbling, doing magic, and the like. No doubt some also did tight-rope walking, as the balancing using a long pole is similar to the rocking of the eagles in flight.

Terathopis has two interpretations of its origin: one is from the Greek *teras* and *opos* for 'meteor' and 'appearing', i.e. appearing meteor-like, and the other also from the Greek *teras* and *ops* for 'face' and 'marvellous'. *ecaudatus* is from the Latin for 'without' and 'tail'.

Measurements
Length: 60–62 cm (61 cm)
Wingspread: 186–190 cm (188 cm)
Weight: 1.8–2.9 kg (2.4 kg)

Harriers—genus *Circus*

Four species of harrier in the genus *Circus* occur in the Western Palearctic. World-wide are 13 species. All are inhabitants of open areas and are characterized by long wings, tails, and legs, and acute hearing used for hunting. Hunting flights are distinctive, a slow quartering flight low over the ground on wings that are held somewhat above the horizontal. When prey is sighted or heard, they brake suddenly, wheel, and pounce. Seen close, most harriers show an owl-like facial ruff; this is thought to be of specialized feathers to direct sound into the ear.

Adults of each sex have distinctive plumages; females are somberly coloured for camouflage while incubating on ground nest. Males, on the other hand, are grey and white; coloration thought to make them more difficult to see against the sky by alert prey (e.g. birds, lizards). Juvenile plumages are similar to those of adult females. Males are smaller, fly faster when hunting, and take proportionally more birds than do females. Wings of males appear relatively narrower than those of females. Adult females and juveniles of the three smaller species are often called 'ring tails', apparently for their banded tails.

Harriers are usually most active during the crepuscular periods but may be seen hunting at any time. Display flights of breeding males are usually spectacular. All species form communal night roosts in grassy areas, often with other harrier species and particularly in winter.

'Harrier' comes from the Old English *hergian*, meaning 'to harass by hostile attacks'. *Circus* is from the Greek *kirkos*, 'circle,' for the bird's habit of flying in circles.

MARSH HARRIER   *Circus aeruginosus*
(Western Marsh Harrier)

Plates 16–18; Photos, p. 310

Description
The Western Palearctic's *largest and darkest harrier*. It has relatively wider wings than other harriers and *lacks white uppertail coverts* in all plumages. Adults have sexually dimorphic plumages; females are noticeably larger than males and have proportionally wider wings. Second winter and third winter males are recognizable. Juvenile plumage is alike for the sexes and is similar to that of the adult female. A rare and local dark-morph plumage exists for each age/sex classes. On perched birds, wingtips fall short of tail tip. Cere is pale greenish-yellow; legs are yellow.

Adult male's head is usually rufous-buffy with a variable amount of narrow dark brown streaking, black lores, and a noticeable owl-like facial disk. Heads of darker birds are blackish-brown with rufous-buffy streaking. *Tricoloured pattern on uppersides* consists of dark brown back and secondary upperwing coverts, silvery primary coverts, secondaries, and inner primaries, and black outer primaries. The uppertail is silvery-grey and unbanded. Uppertail coverts vary from solid grey to grey and rufous to solid rufous. Underwing coverts are whitish, sometimes with faint rufous wash. Secondaries and primaries below appear mostly white with wide black tips on the outer primaries. Underparts consist of a buffy to rufous breast that is heavily streaked with dark brown, and solid rufous belly, leg feathers, and undertail coverts. Undertail appears white. Plumages vary from pale to somewhat dark, but all birds have the same basic pattern. Iris is yellow with an orangish tinge.

Third winter male is similar to adult male, but underwing coverts have heavy rufous wash and dark brown shaft streaking, secondaries have wide dusky tips forming band on trailing edge of wing, and tail has black subterminal band.

Second winter male plumage is similar to that of the adult female, but several features distinguish it: breast is streaked; undersides of secondaries are pale, lacking contrast with pale primaries; uppersides of outer secondaries, inner primaries, and primary upperwing coverts have a silvery cast; and greyish to rufous tail shows a dark subterminal band. Tail can show multiple dark bands like those of some adult females.

Adult female's head is creamy to white with wide dark brown eye-lines that extend from each eye to the shoulder and some fine dark brown streaking on rear crown and nape. Upperparts are dark brown and usually show large creamy patches with fine dark brown streaking on the leading edge of the upperwing coverts, sometimes on the back. Uppersides of flight feathers are dark brown. Tail is dark brown, but inner webs of the outer feathers have rufous mottling; with backlighting or in good light *spread tail appears rufous*. Tail can show multiple dark bands. Greater uppertail coverts are mostly rufous. *Dark brown breast usually is crossed by a noticeable pale band*; belly, undertail coverts, and leg feathers are dark brown with a strong rufous cast. Dark brown underwing coverts usually have some creamy feathers, most often near the leading edge, but on some birds it is extensive. Primaries below are paler than secondaries and have darker tips; this is variable but is usually visible as pale primary wing panels that are not as large nor as white as those of the adult male; panels may have a rufous cast. Secondaries are darker than primaries but usually a shade or two paler than the underwing coverts. Iris is yellow.

Juvenile plumage is *almost uniformly dark brown*, but with creamy to whitish unstreaked crown and throat patches; some birds have adult-female-like creamy patches on the breast or upperwing or underwing coverts. Some birds are completely dark brown, lacking any creamy patches; others have only a creamy unstreaked patch on the nape. Underwing usually shows small white areas at base of primaries but not large pale patches like those of the adult female. Uppertail coverts are dark brown but may be rufous in spring due to moult. Tail is dark brown without rufous mottling (but replacement feathers may have rufous mottling). Iris is dark brown. Males have yellow eyes by their first spring, but females may retain dark coloured eyes into adult plumage. (*Note*: Juveniles on spring migration have already begun moult into adult (female) or subadult (male) plumage.

Dark-morph

Dark-morph plumages exist for all age–sex classes.

Adult male dark morph has *solid dark blackish-brown to jet black head, body, and wing coverts, a whitish wing panel bordered by black on the underwings, and a faint silvery cast to the primary upperwing coverts.* Like the light-morph adult male, iris is yellow with an orangish tinge and tail is unbanded silver-grey.

Second winter male dark morph also has solid dark brown body and wing coverts, but there may be some creamy streaking on throat, crown, and upperwing coverts. Underwing is like that of light-morph first summer male, with a variable amount of whitish speckling in primaries and secondaries. Tail is usually dark brown with a greyish cast but may have faint rufous cast and show faint banding. Iris is yellow.

Adult female dark morph is solidly dark brown overall but with (usually) a streaked creamy nape patch and pale patches on the undersides of the primaries. Like the light-morph adult female, they have yellow eyes and dark brown tails with rufous mottling.

Juveniles in dark-morph plumage are most likely the all-dark juveniles with or without light nape patches as described above.

Unusual plumages

Partial albinos has been reported.

Similar species

Black Kites (Plates 8, 9) appear similar in outline and coloration but are larger, have forked, banded tails (not always visible), and have noticeable barring in the pale wing panels. They also have pale bars on the upperwing coverts—a feature not seen on Marsh Harriers. Kites fly on cupped wings and appear quite unlike harriers, which fly with wings in a dihedral.

Dark-morph Booted Eagles (Plate 37) also have a similar outline and coloration to adult female and juvenile Marsh Harriers, but their pale wing panels are restricted to the inner primaries and they show a dusky tip on undertail. They also show a pale bar on the upperwing coverts, white 'landing lights' at the base of the leading edge of the wings, and a white 'U' above the base of the uppertail—features not seen on Marsh Harriers. They fly on cupped wings and appear quite unlike harriers, which fly with wings in a dihedral.

Dark-morph Montagu's Harrier (Plate 18) are similar in all plumages to dark-morph Marsh Harriers but are smaller and slimmer-winged. Adult males show dark undersides of primaries. Adult females and juveniles show banding in tails.

Harriers (Plates 19–21) of other species in light morph all have white uppertail coverts.

Flight

Hunting flight is typical of harriers. Powered flight is with heavy, slow wingbeats. Soars with wings in a medium dihedral; glides usually with wings held with less dihedral than other harriers, often held in a modified dihedral.

Moult

Annual moult of adults is complete. Migrants begin moult in summer, but suspend it for migration and complete it on winter grounds. Juveniles begin body moult on the wintering grounds, occasionally replacing central tail feathers by spring migration. Some juvenile females can appear adult-like during spring migration. Juvenile moult is completed during the first summer.

Behaviour

Hunt in typical harrier fashion, quartering slowly over preferred habitat of reed beds searching for birds or mammals. While hunting, they will sometimes stop and hover with legs dangling for a second or more. Range of prey is large and includes most animals that occur in reed beds, including fish. In winter they also eat carrion. Often the victims of piracy by eagles and other large raptors, they also pirate prey from other harriers. Favourite perches are in high grass or reeds or on open ground devoid of vegetation. Like other harriers, they form communal night roosts, usually in reed beds.

Display flight of male is a series of undulations but also circling on high. Polygamy, with one male having more than one female, occurs. Marsh Harriers are usually silent, except during breeding.

Status and distribution

Breed locally throughout much of western Europe, but formerly much more common. They are still fairly common in eastern Europe. Drainage of marshes is the most likely cause of any population reduction. They are rare and local in north-western Africa, Turkey, England, Scandinavia, and the Middle East and absent from the far north. Birds from north and east winter in Africa, while the rest stay in areas with milder winter conditions, particularly in southern and western Europe and the Middle East. North African birds are resident.

Dark-morph birds are seen in small numbers regularly on migration and occasionally in winter in Israel, where three birds in this plumage were captured for ringing. They have been reported from various parts of Europe. There is a dark-morph adult male specimen from England.

Fine points

Adult's uppertail coverts are mostly rufous (can be grey on some adult males) and appear paler than the adjacent dark brown back feathers; this can be seen when birds are viewed close in good light. Juveniles have dark brown uppertail coverts.

Subspecies

Two races occur in the Western Palearctic: nominate *aeruginosus* in all areas except north-western Africa, where it is replaced by *harterti*, somewhat smaller.

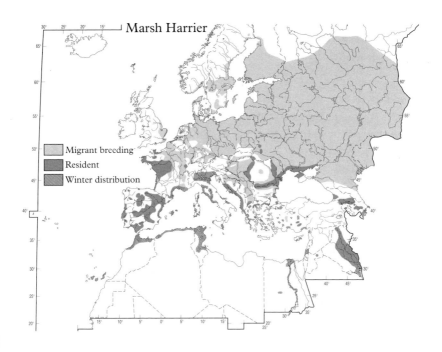

Etymology

aeruginosus is Latin for 'rusty-coloured', in reference to adult's belly and leg feather colours.

Measurements

Length: 42–53 cm (47 cm)
Wingspread: 115–139 cm (128 cm)
Weight: 375–755 g (540 g)

Hen Harrier   *Circus cyaneus*

Plate 19; Photos, p. 312

Description

Medium-sized harrier, the most northern of the Western Palaearctic's harriers. They have white uppertail coverts in all plumages. Adults have sexually dimorphic plumages; females are noticeably larger than males and have proportionally wider wings. First summer males are recognizable. Juvenile plumages of the sexes are almost alike and are very similar to that of the adult female. On perched birds, wingtips fall short of tail tip. Cere is greenish-yellow to yellow, legs are orange-yellow, and beak is black.

Adult male. Head, back, and upperwing coverts are pale grey to grey. Brownish nape patch is present on birds of all ages. Outer six primaries are black; inner four and secondaries are pale grey to grey with black tips on inner webs, latter forming *black band on trailing edge of underwings*. Silvery-grey uppertail shows faint banding. Pale grey to grey breast contrasts with unmarked white belly, leg feathers, and undertail coverts, forming noticeable hood. Undertail appears white, with faint dark banding. Iris colour is orangish-yellow.

Second winter male plumage is similar to that of the adult male, but back is contrastingly darker brown-grey and breast, belly, and flanks may have some fine grey or rufous streaking. Central tail feathers often have dusky tips.

Adult female. Brown head shows buffy-rufous streaking on crown, nape, and cheeks and owl-like facial disk formed by pale tips of facial ruff. Back and upperwing coverts are brown; pale nape streaking may extend on to upper back. Median coverts have pale feather edging; this results in pale, kite-like bars across each upperwing. Uppersides of brown flight feathers show dark banding when seen close. White uppertail coverts form large patch. Upperside of folded tail (central tail feathers) shows even-width brown and dark brown bands and often has a greyish cast; outer tail feathers have more bands of brown and rufous-buff. Buffy underparts are heavily marked with thick brown streaks on breast and with smaller rufous-

brown blobs on belly, leg feathers, and undertail coverts. Underwing shows pale primaries with narrow dark banding, dark secondaries with two pale bands that extend to body, and darkly marked median and greater coverts; lesser coverts are buffy and lightly marked. Iris colour varies with age; on youngest (1 to 2 year olds) it is yellow with numerous brown flecks and appears brownish; flecking is gradually reduced until eyes become clear yellow by 4 to 6 years.

Juvenile plumage is very similar to that of adult female; differences are subtle. Cheek patches show less buffy streaking and appear darker than those of adult female. Breast, belly, leg feathers and undertail coverts average a more rufous-buffy base colour and are marked less heavily with narrower streaking (rather than dark blobs) than are those of adult female, particularly with finer, narrower streaks on leg feathers and undertail coverts. Dark secondaries have pale greyish bands, less noticeable than whitish bands of adult female; bands of males are narrower and extend to body, while those of females are darker, with less contrast, and do not extend to body. Upperside of tail of juvenile female lacks greyish cast. Iris colour also differs, with males having pale grey to grey-brown eyes and females having chocolate brown eyes. Male eye colour becomes yellow by spring.

Note: Juveniles with unstreaked, brighter rufous underparts have been reported from Britain and Holland, possible vagrants from North America. There is also a report of an adult male in Holland that resembled the North American race.

Unusual plumages

Albinism, partial albinism, and even melanism, have been reported for this species. See Watson (1977, p. 42) for drawings of several partial albinos.

Similar species

Montagu's Harriers (Plates 20, 21) appear similar. See under that species for distinctions.

Pallid Harriers (Plates 20, 21) appear similar. See under that species for distinctions.

Marsh Harriers (Plates 16, 17) are larger and overall darker, with wider wings, and never show white patch on uppertail coverts.

Flight

Hunting flight is typical of harriers. Powered flight is with slow wing-beats on flexible wings. Soars with wings in a slight dihedral; glides usually with wings in medium dihedral, but often in a modified dihedral.

Moult

Annual moult of adults is complete. Juveniles begin body moult on the wintering grounds, but begin replacing flight and tail feathers in late spring and summer, completing in autumn or winter.

Behaviour

Hunt in typical harrier fashion, quartering slowly over various types of open terrain searching for birds or mammals. Only rarely are reptiles, insects, or birds' eggs taken. While hunting, they will sometimes stop and hover with legs dangling for a second or more. Favourite perches include small bushes, mounds, fence posts, stone walls, and ground.

Display flight of male is a series of undulations but also loop-the-loops and circling on high. They nest on ground in various open habitats, including grasslands, upland moors, heath, open taiga, forest clearings, and even conifer plantations but have been reported nesting in trees in Northern Ireland. Polygyny, with one male having more than one female, occurs regularly. Hen Harriers are usually silent, except during breeding.

Like other harriers, they form communal night roosts, especially in winter and usually in grass or reeds.

Status and distribution

Breed throughout much of northern and eastern Europe, farther north than congeners. They are also sedentary local breeders in south-west and central Europe and Britain. They are less restricted to wet areas than are Marsh and Montagu's Harriers.

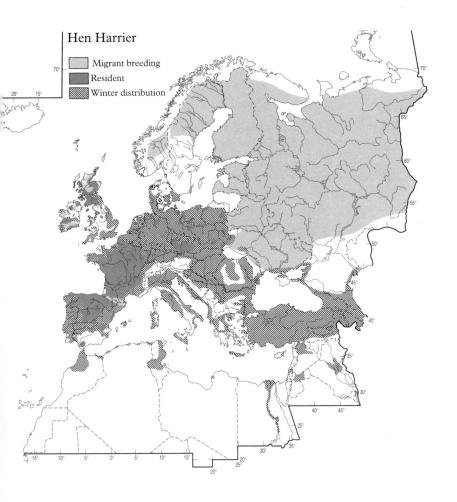

Hen Harrier

- Migrant breeding
- Resident
- Winter distribution

Brids in north and north-east are migratory and move to south-ern Europe for winter, barely into north Africa and the Middle East. In winter they utilize a wider variety of open habitats.

Fine points

Adult female and juvenile Hen Harriers have two pale bands across dark secondaries on underwing; Montagu's and Pallid Harrier ring tails have only one. Rare melanistic birds lack white uppertail coverts.

Subspecies

Circus c. cyaneus occurs throughout the Western Palearctic. *C. c. hudsonius* is a vagrant from North America and could be a separate species.

Etymology

cyaneus comes from Greek *Kyaneous*, 'dark blue', for male's back colour.

Measurements

Length: 41–51 cm (46 cm)
Wingspread: 99–122 cm (110 cm)
Weight: male 290–400 g (350 g)
        female 410–600 g (520 g)

Montagu's Harrier   *Circus pygargus*

Plates 18, 20, 21; Photos, p. 314

Description

One of two slender pointed-wingtip harriers breeding in the Western Palearctic, a summer resident throughout the middle latitudes of Europe. Plumage of adult male is distinctive from adult male of the other small harrier, Pallid Harrier, but adult females and juveniles are distinguished in the field only by careful scrutiny. Adults have sexually dimorphic plumages; females are somewhat larger, but there is size overlap. Juvenile plumage is somewhat similar to that of adult female. *On perched birds, wingtips reach or nearly reach tail tip.* An uncommon and local dark colour morph exists for each age and sex class.

Adult male head and breast are medium grey, with latter usually forming bib. Facial ruff is not obvious. *Belly, legs, and undertail coverts are usually white with narrow reddish streaks but are sometimes solid pale grey*; birds so marked lack bibs and hooded appearance. Back and upperwing coverts are darker grey; latter often have black spots. Outer primaries are black; *inner primaries and secondaries are pale grey with one or two black bands crossing every feather.* White underwing coverts are heavily marked with large rufous spots. Uppertail coverts are white with grey spots; a white patch is usually not discernible. Upperside of central tail feathers are pale grey; other tail feathers are grey on outer webs and whitish with wide rufous to brown bands on inner webs. Iris, cere, and legs are bright yellow.

Second winter males are similar to adult males, but are darker overall, with thicker rufous streaking on belly, legs, and undertail coverts and usually show retained juvenile feathers, particularly outer primaries. Central grey tail feathers have a dark subterminal band.

Adult female. Brown head has small white spots above and below, a large dark brown patch below and behind each eye, and an owl-like facial disk, which *does not extend across throat.* Crown and nape show some narrow rufous-buffy streaking. Back and upperwing coverts are brown; upper back often has some rufous-buffy streaking. Median upperwing coverts show some pale feather edging; this results in a pale kite-like bar across each upperwing. Flight feathers above are same colour as coverts, but darker brown banding is visible on close birds. White uppertail coverts have dark brown spotting or barring but are noticeable as a small white patch at base of tail. Creamy to whitish underparts are heavily streaked dark brown on breast; streaking becomes lighter and more rufous on belly, leg feathers, and undertail coverts. Underwing coverts are rufous-brown with pale edges. Primaries below are whitish with several rows of distinct narrow dark brown bands. Secondaries are brown with two noticeable white bands across each feather; *widest band extends to body*. Central pair of tail feathers above are brown, with indistinct darker bands; others are brown with wide paler bands except for outer ones, which are banded rufous-buff and white. Iris, cere, and legs are yellow, duller than the colour of adult male.

Second winter female. Similar to adult female but with brown eyes, rufous underparts and underwing coverts, and dark uppersides of secondaries, such that dark bands not noticeable.

Juveniles are similar to adult female, but can be easily separated in the field by uniformly dark secondaries on underwing and unmarked rufous underparts (many birds show faint narrow darker rufous shaft streaks on upper breast and flanks). Back and upperwing and uppertail coverts have rufous-buffy tips in fresh plumage. Pale tail bands are rufous-buff. Juvenile males and females differ in iris colour and pattern of undersides of outer primaries. Juvenile males from below show whitish outer primaries with dark mottling (usually no banding, but sometimes bands at base of primaries), and females show banded pattern like that of adult females, but with a rufous wash. Male iris is grey to grey-brown; female's is dark brown.

Juveniles (second summer) returning in spring can appear quite adult-like because of body moult during winter, but amount of moult is variable.

Dark-morph

Dark-morph plumages exist for all age–sex classes. Uppertail coverts on dark-morph birds may be pale but are not white.

Adult male dark-morph is pale (rarely) to medium (sometimes) to dark (usual) sooty-grey overall except for black primaries and grey secondaries, which lack the black bands of light-morph birds. Grey tail is unbanded and can appear whitish below.

Second winter males differ from adult males in being overall dark sooty-brown including dark brown secondaries and underwing coverts. Noticeable whitish outer primaries may be retained during transition from juvenile plumage.

Adult female dark-morph is uniform dark brown except for rufous-buff nape patch and undersides of flight and tail feathers; primaries are whitish and banded as in light morph and secondaries are dark brown with a hint of pale bands of light morph. Uppertail is dark brown with some banding on outer feathers.

Juveniles are similar to dark-morph adult female but have uniformly dark secondaries and more distinct rufous tail banding on outer tail feathers. Sex differences in underside of primaries and iris colour are same as these of normal-morph juvenile.

Unusual plumages

Juvenile hybrids with Pallid Harrier resemble juvenile Pallids.

Similar species

Pallid Harriers (Plates 20, 21), especially adult females and juveniles, are very similar to similar-age Montagu's. Adult male Pallid is much paler grey, lacks dark bands in secondaries on upper and lower wings and hooded appearance (except for subadult), and has smaller, more wedge-shaped black areas on outer primaries. Adult female Pallid has less distinct wide pale band on secondaries that does not extend to body and no visible banding on upperside of flight feathers. More distinct pale ring around face that extends across throat is noticeable on close Pallids. Juvenile Pallid is distinguished complete pale ring around head and lack of flank streaks. Perched Pallids show longer legs and wing tips that fall somewhat short of tail tip.

Hen Harrier (Plate 19) adult females and juveniles are similar to adult female Montagu's, but notice Hen Harrier's wider wings, especially more rounded wingtips, larger size, and larger white patch at base of uppertail. Adult males are a paler, cleaner grey and show white patch on uppertail coverts.

Dark-morph Marsh Harriers (Plate 18) are also overall dark like dark-morph Montagu's. See under that species for distinctions.

Flight

Hunting flight is typical of harriers. Powered flight is light and buoyant with slow wingbeats of flexible wings. Soars with wings in a strong dihedral; glides with less of a dihedral or sometimes with a modified dihedral.

Moult

Annual moult of adults is complete and begins in early summer but is suspended during breeding and resumes when young are almost independent until migration begins, when it is suspended again. It is completed on wintering grounds in Africa. Juveniles begin body moult on wintering grounds, and birds (particularly males) returning in spring show a range of variation from some that appear very adult-like to others that are essentially paler (faded) versions of juveniles. Usually flight feathers are not replaced until summer, but some tail feathers, especially central ones, may be replaced during winter. Juvenile moult is completed during the first summer.

Behaviour

Hunt in typical harrier fashion by quartering slowly over fields. Upon locating prey, they wheel and pounce quickly to ground. They are partial to cultivated areas during summer, often nesting on ground in wheat fields, but also hunt and nest in other open areas, such as bushy thickets, even young conifer plantations. They eat mainly small mammals, birds, and reptiles, sometimes insects, particularly in winter, and on occasion, birds' eggs.

Display flight of male is a series of undulations, and sometimes includes some loop-the-loops and vertical plunges. Polygyny, with one male having more than one female, occurs occasionally. Montagu's are usually silent, except for during breeding.

Like other harriers, they form communal night roosts, particularly during winter, but also in early autumn.

Status and distribution

Breed in the Western Palearctic from Spain and France locally through central Europe north to 60°N and east into the former USSR and western Asia but are rare and local in England, Turkey, and north-western Africa. They do not breed north of 60°, in the Middle East, nor in the rest of North Africa. Entire population migrates into sub-Saharan Africa for winter, but a scattering of winter records exist.

Dark-morph birds are seen most often in Spain, with breeding records for eastern Europe and sight records from France, Italy, and England. They have not been recorded in Middle East.

Montagu's Harrier

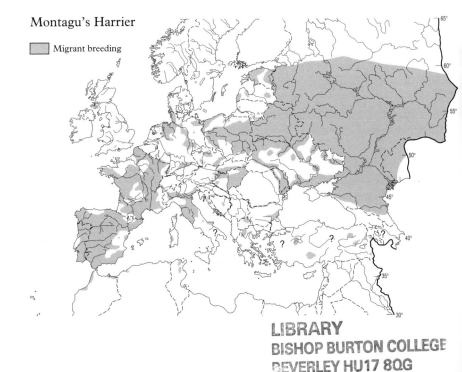

Migrant breeding

Fine points

Small indistinct supraorbital ridge gives Montagu's a pigeon-headed appearance; Pallids have the ridge and appear more raptorial. Emargination of next to outer primary occurs more than two cm beyond tip of primary upperwing coverts; that of Pallid Harriers occurs at tips of coverts.

Subspecies

None.

Etymology

pygargus is from the Greek *puge* and *argos* for 'white rump'.

Measurements

Length: 39–46 cm (43 cm)
Wingspread: 102–116 cm (109 cm)
Weight: male 212–300 g (260 g)
        female 260–435 g (360 g)

Pallid Harrier   *Circus macrourus*

Plates 20, 21; Photos, p. 316

Description

One of two slender pointed-wingtip harriers breeding in our area, a summer resident in east central Europe, occasionally farther north, west, and south. Plumages of adults are different; females are separably larger. Plumage of adult male is distinctive from adult male of the other small harrier, Montagu's Harrier, but adult female and juvenile are distinguished in the field only by careful scrutiny. Juvenile plumage is somewhat similar to that of adult female. On perched birds, wingtips fall short of tail tip by 2–5 cm.

Adult male. Head and underparts are pearly grey. Back and upperwing coverts are light grey; flight feathers are the same light grey, except for *black wedge on outer primaries* (nos. 6–9). Greater uppertail coverts are white with large light grey spots, but no white patch is noticeable. Breast is pale grey and on some birds contrasts somewhat with rest of white underparts, forming a faint bib. *Underwing is completely white except for black wedge* in primaries. Central tail feathers are light grey; other tail feathers are progressively whiter outward with light grey banding. Outer pair are white with indistinct rufous banding. Iris is orangish-yellow, cere is yellow, and legs are orangish-yellow.

Second winter male is similar to adult male and shows same black wedges on wings but is a darker, more brownish-grey on head, back, and upperwings and has brownish nape patch. Darker grey breast contrasts with whitish belly forming a noticeable dark bib; usually has noticeable rufous to dark brown streaking on breast and belly. Dusky terminal bands on undersides of secondaries form a dusky band on trailing edge of wing. Grey central tail feathers show hints of dark bands on uppersides; whitish outer tail feathers have rufous and grey banding. Undertail coverts often show sparse greyish or rufous markings.

Adult female. Brown head has a small white spot above and below each eye, with narrow rufous-buffy streaking on crown and nape. Pale tips on owl-like facial ruff form pale collar around face, *including the throat.* Back and upperwing coverts are brown; upper back shows some rufous-buffy streaking. Median upperwing coverts show some pale feather edging; this, results in a pale kite-like bar across each upperwing. White uppertail coverts have dark brown barring or spotting but are noticeable as a small white patch at base of tail. Creamy to whitish underparts are heavily streaked dark brown on breast; streaking becomes narrower and more rufous on belly, leg feathers, and undertail coverts. Underwing coverts are creamy to whitish and streaked with rufous-brown, heaviest on median coverts and axillaries. *Primaries below usually lack dark tips* and are whitish with several rows of distinct dark bands. Secondaries are dark brown with a noticeable wide white band on outer ones, but *band does not extend to body.* Central pair of tail feathers are dark brown, with even darker indistinct bands; others are brown with wide rufous-buffy bands. Iris, cere, and legs are yellow, duller than those of adult male.

Juveniles are similar to adult female, but can be easily separated in the field by unmarked rufous underparts and dark undersides of secondaries. Face shows *wide pale collar that extends across nape.* White uppertail coverts are less heavily marked. Juvenile males and females differ in iris colour. Male iris is grey to grey-brown and yellowish by spring; female's is dark brown becoming medium brown by spring.

Unusual plumages

Juvenile hybrids with Montagu's Harrier resemble juvenile Pallids. A specimen of a second winter male was unusual in that it had grey-brown upperparts, adult female-like white face patches, and heavy brown streaking on breast, as well as less intense rufous markings on belly, leg feathers, and undertail coverts. Black wedges on outer primaries were dusky and ill defined. A dilute-plumage juvenile was photographed in India.

Similar species

Montagu's Harrier (Plates 20, 21) in light-morph plumages, especially adult females and juveniles, are very similar to Pallids. See under that species for distinctions.

Hen Harrier (Plate 19) adult females and juveniles are similar to adult female Pallid, but notice wider wings, especially more rounded wingtips, larger size, and larger white patch at base of uppertail. Adult male Hens have grey bibs and black band on trailing edge of wings similar to those of subadult male Pallid, but black on wingtips is more extensive and not wedge-shaped and white patch on uppertail coverts is noticeable.

Flight

Hunting flight is typical of harriers but usually faster than that of Montagu's Harriers. Powered flight is light and buoyant with slow wingbeats of flexible wings. Soars with wings in a strong dihedral; glides with less of a dihedral or sometimes with a modified dihedral.

Moult

Annual moult of adults is complete and begins in early summer but is suspended during breeding and continues from when young are almost independent until migration begins, when it is suspended again. It is completed on wintering grounds. Juveniles begin moult after their return in spring to the breeding grounds and complete it before autumn migration.

Behaviour

Hunt in typical harrier fashion by quartering slowly over fields. Upon locating prey, they wheel and pounce quickly to ground. They prefer dryer areas than do other harriers, particularly grasslands. They eat mainly small birds, but also mammals, reptiles, sometimes insects, and on occasion, birds' eggs.

Display flight of male is a series of undulations, and sometimes includes some loop-the-loops and vertical plunges. Polygyny, with one male having more than one female, has not been reported for this species. Pallids are usually silent, except during breeding.

Like other harriers, they form communal night roosts, particularly during winter, but also in early autumn.

Status and distribution

Breed in the Western Palearctic mainly in the former USSR; however, sporadic breeding farther north, west, and south in Sweden, Germany, Romania, and Turkey has been reported. An adult male in May in the Netherlands was displaying and offering food to an adult female Hen Harrier. Entire population migrates with most going into sub-Saharan Africa, but many winter records, particularly in the Middle East but also southern Europe, exist. Autumn migrations are primarily through the eastern Mediterranean. Return flights north in spring occur farther west-ward; many are seen at the Straits of Messina. Vagrants have been reported from almost everywhere in Europe.

Pallid Harrier

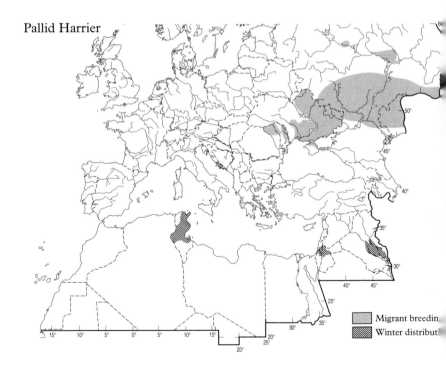

Migrant breedin

Winter distribut

Fine points

Emargination of next to outer primary occurs at tips of primary upperwing coverts; that of Montagu's Harriers occurs more than 2 cm beyond tips of these coverts.

Subspecies

Monotypic.

Etymology

macrourus is from Greek *makros* for 'tail' and *kirkos* for 'long'.

Measurements

Length: 41–49 cm (44 cm)
Wingspread: 102–120 cm (109 cm)
Weight: male 235–385 g (310 g)
          female 380–520 g (440 g)

CHANTING GOSHAWKS—GENUS *MELIERAX*

Three species of Chanting Goshawks occur in sub-Saharan Africa. All are characterized by brightly coloured faces and legs, long legs, and long tails.

One, the Dark Chanting Goshawk, occurs in the Western Palearctic locally in central Morocco, where it is often seen perched conspicuously in open woodlands or thornbush.

Chanting Goshawks are taxonomically close to Accipiters.

Dark Chanting Goshawk *Melierax metabates*

Plate 47

Description

A medium-sized raptor common in sub-Saharan Africa, a rare local resident in central Morocco and vagrant in Spain and Israel. Long legs and tail are distinctive. Sexes are alike in plumage; females are larger. Juveniles differ in plumage from adults.

Adult. Head, upperparts, and breast are dark grey. Belly, leg feathers, underwing coverts, and tail coverts are finely barred white and slate grey. Flight feathers are slate grey above and whitish below, except for whitish marbling on upper secondaries and mostly black outer primaries. Long tail has dark inner feathers with white tips; outer tail feathers have wide equal-width dark and pale bands. Cere, gape, and legs are bright orange-red. Eye is dark brown.

Juvenile. Head is dark brown with pale superciliaries and darkly streaked pale throat. Upperparts are dark brown, with narrow pale edges to upperwing coverts in fresh plumage. Brown breast shows some buffy streaking. Belly, leg feathers, underwing coverts, and tail coverts are barred brown and white. Uppersides of flight feathers are brown; undersides are whitish with many narrow dark bands, except for brown tips of outer primaries. Tail has several wide equal-width dark and paler brown bands; pale bands becoming wider and more whitish on outer feathers. Cere and gape are yellow, and legs are orangish. Eye is yellow to pale brown.

Unusual plumages

No unusual plumages have been reported.

Similar species

With their long red to orange legs and long tails, they are quite distinctive and unlike any other Western Palearctic diurnal raptor.

Flight
Powered flight is direct and quite accipiter-like, with several rapid wingbeats and then a glide. Soars with wings in a dihedral; glides with wings level or slightly cupped.

Moult
Undergo a complete moult annually; like most African raptors, they can be in active moult all year. Post-juvenile moult is also complete.

Behaviour
Somewhat buzzard-like, in that they spend much of their time perched in conspicuous places, such as pole or tree tops, and often capture prey on the ground after a short flight. However, they can also be rather accipiter-like in their flights after agile prey. They also hunt regularly by walking on the ground. They take a wide variety of prey, including lizards, snakes, birds, mammals, and insects.

Adult pairs are territorial and build stick nests in the centre of the crown of acacia or similar trees, often incorporating non-stick objects, such as dung and rags, into the nest structure. They usually raise only one young, although two eggs are often laid.

Vocalizations for which they are named are usually only given during the breeding season.

Status and distribution
Only a dozen or so pairs of Dark Chanting Goshawks breed in the Western Palearctic, most, if not all of them in the Plain of Sous in central Morocco. A vagrant was recorded in Israel. A sight record from Spain is not convincing.

Fine points
Two similar species, Eastern Chanting Goshawk and Pale Chanting Goshawk, occur in east and southern Africa, respectively. Juveniles of the three species are most difficult to separate in the field.

Dark Chanting Goshawk

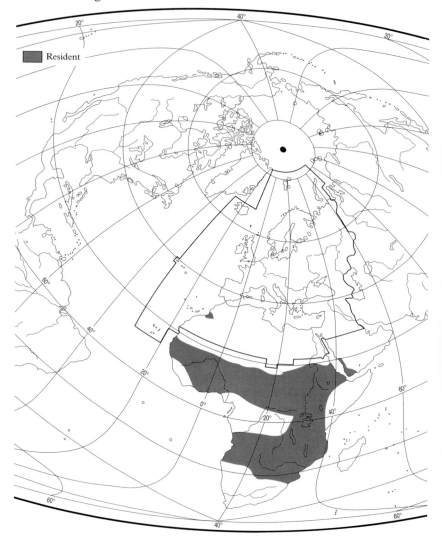

Resident

Subspecies

The subspecies *theresae* occurs in Morocco. However, the races are not well differentiated.

Etymology

Melierax is from the Greek *melos* and *hierax* for 'a song' and 'a hawk'; *metabates* is from the Greek *metabasis* meaning 'different from', presumably from the other Chanting Goshawks.

Measurements

Length: 43–56 cm (50 cm)
Wingspread: 92–109 cm (101 cm)
Weight: 450–850 g (650 g)

HAWKS—GENUS *ACCIPITER*

Accipiters are short-winged, long-tailed, forest-dwelling raptors. Three species occur regularly in the Western Palearctic; another occurs as a rare vagrant. Hawks are aggressive, capable of rapid acceleration, and relentless and reckless in pursuit of prey. Sexual dimorphism, with females much larger than males, is pronounced in this genus.

Because of their forest preference, especially for nesting, accipiters are less visible than are other raptors. During migration, they may be more conspicuous, particularly at migration concentration locations, especially in Israel.

All four species soar regularly, usually for a while every day. Most of their migrations are accomplished by soaring.

Accipiter is Latin for 'bird of prey'. It probably derived from *accipere*, meaning 'to take', but another possible origin is from the Greek *aci* for 'swift' and *pertrum* for 'wing'. 'Hawk' comes from the old Teutonic root *haf* or *hab*, meaning 'to seize'. 'Hawk' properly refers only to raptors in this genus, but sometimes it is used generically to refer to all diurnal raptors.

Shikra   *Accipiter badius*

Plate 22

Description

This small accipiter is a rare breeder in the Caucasus and a rare vagrant to the Middle East. Adult plumages are similar; juvenile plumage is different from those of adults. Females are separably larger. Cere is greenish-yellow, beak is dark, and legs are yellow.

Adult male. Head is grey except for creamy throat with a narrow blackish mesial stripe. Small patches of rufous extend from upper breast to each side of lower neck. Back, upperwing coverts, and uppersides of flight feathers are grey, shading to a paler grey on rump and uppertail coverts. Grey upperside (central pair) and whitish underside (outer pair) of folded long grey tail show no bands; fanned tail shows narrow dark bands, with subterminal band wider than others. Creamy underparts are heavily but finely barred rufous, often appearing solid rufous on distant birds. Creamy leg feathers are barred rufous. Creamy undertail coverts are usually unmarked. Creamy underwings show unmarked coverts and lightly barred flight feathers with dark tips. Iris varies from orange to red.

Adult female is similar in plumage to adult male but back is usually a darker grey often with a brownish cast and rufous barring on underparts is heavier and courser. Wingtips do not appear as dark as those of male. Central pair of grey tail feathers have a dark subterminal band and often show several dark spots along vanes. Iris varies from yellow to orange.

Juvenile. Head is brown except for short creamy superciliaries in front of and over eyes and creamy throat with narrow dark rufous to dark brown mesial stripe. Up close some fine creamy and rufous streaking on crown, nape, and cheeks are noticeable. Back and upperwing coverts are brown; many feathers have rufous edges. Uppersides of flight feathers are brown with narrow darker bands. Upperside of long brown tail shows four narrow dark brown bands. Creamy underparts are marked rufous to dark brown, streaked on breast, spotted on belly, and barred on flanks. Creamy leg feathers are marked with small rufous to dark brown spots. Creamy under-wings show sparse dark spots and streaks on coverts and noticeable narrow dark bands on flight feathers. Creamy undertail coverts are usually unmarked. Pale underside of tail shows four or five dark bands, except for outer feathers, which show more than six narrower dark bands. Iris is pale yellow to yellow.

Unusual plumages
No unusual plumages have been reported.

Similar species
Levant Sparrowhawk (Plate 22) is similar in plumage and size but has much more pointed wingtips. Adult Levants have darker, more blue-grey backs and upperwing coverts, more heavily barred leg feathers, and dark-brown irides. Juvenile Levants compared to juvenile Shikras are somewhat darker brown on back and upperwing coverts, show a pale nape patch and heavier markings on underwing coverts, barring on undertail coverts, and lack bold barring on flanks. They show five or more dark bands on uppertails. Both species can show dark throat stripes.
Eurasian Sparrowhawk (Plate 23) is similar in size and plumage, but lacks dark throat stripe. Adults are darker, more blue-grey on back and upperwing coverts. Adult females and juveniles have dark barring on underparts. Eurasian Sparrowhawks show square corners on tip of tail; those of Shikras are somewhat rounded.

Flight
Active flight is typical accipiter, with three to five rapid wingbeats interspersed with a short glide; wingbeats are quicker, more frantic, than those of Sparrowhawk. Soars and glides on flat wings, but sometimes glides with wingtips slightly down.

Moult
Annual moult of adults is completed during the summer. Juveniles' moult is also complete and begins before that of adults.

Shikra

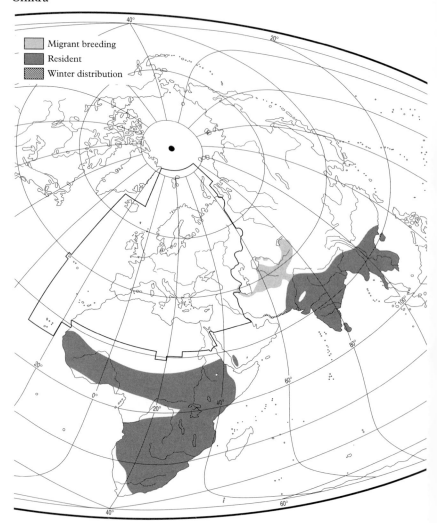

Migrant breeding
Resident
Winter distribution

Behaviour

Typical still hunters that hide in inconspicuous perches, usually in trees, and make short rapid dashes to snatch prey. Their main prey are lizards and insects, but a wide variety of other prey are taken, including small mammals, birds, snakes, and frogs. They have been seen to enter buildings to snatch geckos from walls.

Their nests are built in trees. Spectacular displays include sky dancing and undulating flight, with both sexes participating. Shikras are quite vocal, particularly while breeding.

Status and distribution

Rare summer residents of extreme southern Armenia and southern and eastern Azerbaijan. Birds from this area and northern Iran are migratory, with most moving south-east into Pakistan and India for winter, but a few may move directly south into southern Arabia or east Africa (an adult female was captured and ringed at Elat, Israel in April, presumably returning from one of these areas).

Fine points

Wingtips of Shikra and Sparrowhawk are both rounded, but wingtip of Shikra consists of only five fingered primaries, whereas that of Sparrowhawk has six.

Subspecies

Accipiter badius cenchroides breeds in Caucasus and northern Iran. Apparently some migrate south into the Arabian peninsula.

Etymology

'Shikra' comes from the Hindi *shikari* meaning 'hunter'; *badius* is Latin for 'brown'.

Measurements

Length: 30–36 cm (33 cm)
Wingspread: 64–74 cm (69 cm)
Weight: 140–260 g (210 g)

Levant Sparrowhawk   *Accipiter brevipes*

Plate 22; Photos, p. 318

Description

This small accipiter is a summer breeder in south-eastern Europe. They are gregarious on migration. In flight, their pointed wingtips give them a somewhat falcon-like appearance. Adult plumages are sexually dimorphic; females are somewhat larger. Juvenile plumage is quite different from that of adults. Second summer birds are juveniles returning in spring that have undergone a more or less complete body moult.

Adults

Adults. Cere is yellow, legs are orange-yellow, and iris is dark brown, often with a dark reddish cast, rarely red.

Adult male. Head is blue-grey except for white throat that often shows a faint to noticeable dark mesial stripe. Back, upperwing coverts, uppersides of flight feathers, and uppertail coverts are blue-grey. Upperside of tail is blue-grey; folded uppertail (central pair of feathers) is unbanded, whereas spread tail shows noticeable darker bands on other feathers with subterminal band wider than others. Underparts and leg feathers are finely barred white and pale rufous. Undertail coverts are white. Underwings show whitish coverts that are faintly barred with pale pink and whitish flight feathers that show at most faint banding and black tips on outer primaries. Underwings appear whitish with dark tips on distant flying hawks. Folded undertail appears white and is usually unbanded.

Adult Female is similar to adult male in pattern, but head, back, upperside of wings, and uppertail coverts are darker, more slate blue-grey, often with a brownish cast. Underparts and leg feathers are white, boldly marked with a mixture of rufous and dark brown barring. Underwings show heavy dark barring on coverts and dark banding on flight feathers; outer primaries have dark tips. Whitish undertail coverts show widely spaced narrow dark barring. Underside of tail shows dark brown bands.

Juvenile. Dark brown head shows pale superciliaries, pale nape patch, and pale streaking on cheeks. White throat has wide dark mesial stripe. Back and upperwing and uppertail coverts are dark brown with narrow rufous feather edges in fresh plumage. Uppersides of flight feathers are brown with darker brown banding. Uppertail shows four or five sets of even-width dark and pale brown bands, with dark bands sometimes fainter on central pair of feathers. Whitish underparts are heavily marked with dark brown to rufous-brown spade-shaped blobs, forming wide streaks, but with wide dark brown to rufous-brown barring on flanks. Whitish legs and undertail coverts are boldly barred with dark brown. Whitish underwings show lesser coverts marked with small dark brown to rufous-brown spots, other coverts marked with short wide dark bars, and flight feathers marked with five rows of dark brown bands. Wingtips are not dark. Dark bands on outer pair of tail feathers are narrower as seen on underside of folded tail. Cere is pale yellow, legs are yellow, and iris is medium brown.

Second summer. Juveniles returning in spring have undergone a more-or-less complete body moult and may have replaced some flight and tail feathers (see under 'Moult'). Underparts often show a few retained breast feathers; these appear as short dark streaks. Iris is now dark brown and lacks any of the reddish tones of adults.

Unusual plumages
No unusual plumages have been reported.

Similar species
Eurasian Sparrowhawks, particularly adults (Plate 23), can appear similar to adult Levants. See under that species for distinctions.

Lesser Kestrel adult males (Plate 38) also appear white below with dark, pointed wingtips but have two-toned brown upperwings, rufous backs, and a wide black subterminal tail band.

Flight

Active flight is more leisurely and less dashing than that of other hawks, with slower wing beats. Soars and glides on flat wings.

Moult

Annual moult of adults is complete; it is begun on the breeding grounds, suspended during migration, and completed on the winter grounds. Moult of juveniles begins on winter grounds with most or all body and covert feathers being replaced; some juveniles have replaced some flight and tail feathers, most have not.

Behaviour

Unlike typical sparrowhawks, Levant Sparrowhawks do not specialize in hunting small birds. They prey mainly on lizards and large insects, with some small mammals, snakes, frogs, and birds also taken. Prey is usually captured on the ground after a short sally from a perch or from gliding flight.

Nests are built in trees in copses, wood lots, riverine forests, tree belts, and orchards. While breeding, pairs are territorial. Breeding behaviour is little studied, but should be similar to that of the closely related Shikra. They are quite vocal while breeding, but are generally silent at other times. Calls closely resemble those of Shikra and are unlike those of Goshawk and Sparrowhawk.

They are gregarious on migration, with flocks of over 1000 reported in Israel in both autumn and spring. A report of migration at night requires verification.

Status and distribution

A summer resident of south-eastern Europe, breeding only in the Western Palearctic from Albania, the former Yugoslavia, and Greece eastward to western Turkey and north of the Black Sea across Bulgaria, Romania, southern Ukraine, and southern Russia. As they have been little studied, their status is largely unknown. They prefer open woodland, fields, meadows, and steppes with some trees for breeding.

Levant Sparrowhawk

Migrant breeding

The entire population is migratory, with all or almost all passing through Israel twice a year. Migration is compressed in time, with most hawks passing within a 2-week period. Counts of approximately 40 000 are recorded there most years.

They are gregarious on migration and most often seen in large flocks, when they appear somewhat like bee-eaters. On migration they form communal night roosts. Their wintering grounds south of the Sahara are mostly unknown, as there are few records.

Vagrants have been reported from Tunisia, Italy, Austria, the Czech Republic, Slovakia, and Kazakhstan.

Fine points
Many field guides tout the lack of mesial stripe of adult Levant Sparrowhawks as a field mark, but most adults show this feature.

Subspecies
None.

Etymology
brevipes is from the Latin *brevis* for 'short', and *pes* for 'foot', in reference to this species' short toes.

Measurements
Length: 30–36 cm (33 cm)
Wingspread: 64–76 cm (69 cm)
Weight: 135–240 g (184 g)

Eurasian Sparrowhawk   *Accipiter nisus*

Plate 23; Photos, p. 320

Description

A small accipiter that is widespread and common throughout most of Europe; its range extends barely into north-west Africa and the Middle East. Plumages of adults are sexually dimorphic, but with some overlap in characters. Juvenile plumages are similar to that of adult female. Females are separably larger. *Underparts are barred in all plumages*. Long tail is square-tipped, often with tip notched. Fresh tail feathers have narrow white tips. Beak is black, and cere is pale greenish-yellow to yellow. Long stick-like legs are yellow.

Adult male. Head is slate to dark blue-grey except for rufous cheeks and buffy throat and lacks pale superciliary. Back and upperwing coverts are slate to dark blue-grey, shading to paler blue-grey on uppertail coverts. Upperside of long tail appears brown with a blue-grey cast and has four rather indistinct dark bands, but underside appears pale with dark bands. Bands on outer tail feathers are narrower (seen as underside of folded tail of perched birds). Creamy underparts are barred rufous; colour is often solid rufous on sides of breast and flanks. Creamy leg feathers have narrow rufous barring. Creamy undertail coverts are usually unmarked but may be lightly marked with small rufous arrows. Underwings are pale with bold dark barring on flight feathers and fine rufous and dark grey barring on underwing coverts. Iris varies from orange to yellow-orange but can be red on males in extreme east.

Adult female. Head is slate to dark brown, except for creamy superciliary, throat, and cheeks—all of which have heavy narrow dark streaking. Note the grey Goshawk-like eye-lines. Many birds show two white nape spots and a small rufous area at the rear of each cheek. Back and upperwing coverts vary from dark brown to slate, on latter shading to blue-grey on uppertail coverts, some of which usually have narrow white tips. Upperside of long tail appears brown often with a blue-grey cast and has four or five dark brown bands, but underside appears pale with dark bands. Bands on outer tail feathers are narrower (as seen on underside of folded tail of perched birds). Creamy underparts are finely barred with grey brown to dark slate, but undertail coverts are unmarked. Underwings are pale with bold dark barring on flight feathers and fine dark grey barring on underwing coverts. Iris varies from yellow-orange to yellow.

Juvenile male is similar to brown-backed adult female except that upper breast has obvious rufous streaking. Rest of undersides are barred like adult female but also include more rufous barring and some rufous arrowhead-shaped markings. Back and upperwing coverts have wide rufous edges. Pale tips on greater upperwing coverts form a narrow pale line across upperwings in fresh plumaged juveniles. Uppertail coverts have narrow rufous tips. Creamy undertail feathers are usually unmarked. Iris is pale yellow to yellow.

Juvenile female is more like brown-backed adult female and cannot always be differentiated in the field. Main differences are rufous tips on some uppertail coverts and rufous arrowhead markings on underparts of juvenile females. Back and upperwing coverts have narrower rufous edges (compared to those of males). Pale tips on greater upperwing coverts form a narrow pale line across upperwings in fresh plumaged juveniles.

Other races. *A. n. wolterstorffi* is smaller, darker above, and more densely barred below, *A. n. punicus* averages slightly paler and larger, and *A. n. granti* is slightly darker and more densely barred below.

Unusual plumages

Albinism has been reported, and dilute-plumage specimens were taken in Italy, Germany, and Sweden.

Similar species

Northern Goshawk (Plate 23), particularly the smaller adult male, can appear similar to adult female Eurasian Sparrowhawk but is larger, has more tapered wingtips, more robust breast, larger head, longer neck, and tail with rounded or wedge-shaped tip. In all plumages Northern Goshawks have a stronger face pattern. Juvenile Northern Goshawks are always separable by streaked underparts and underwing coverts and extensive pale feather edging on upperparts.

Levant Sparrowhawk (Plate 22) adults are similar in size and plumage to adult Eurasian Sparrowhawks but have pointed wingtips, dark wingtips on underwings, more rounded tail tips, dark mesial throat stripe, dark eyes, and solid grey cheeks. Juvenile Levants are always distinguishable by spotted to streaked underparts, dark mesial throat stripe, dark eyes, and pale nape patch.

Common Kestrels (Plates 38 and 39), surprisingly, can appear quite similar in size and shape to flying Eurasian Sparrowhawks, which often show pointed wingtips in powered flight. But Common Kestrels are readily distinguished by two-toned pattern on uppersides, streaked underparts, and dark moustache marks. Common Kestrels fly with continuous wingbeats, not gliding after three to five beats, as do Eurasian Sparrowhawks.

Merlin (Plate 43) adult male can appear similar in shape and colour to adult male Eurasian Sparrowhawk. See under that species for distinctions.

Flight

Active flight is typical accipiter, with three to five rapid wingbeats interspersed with a short glide. Soars and glides on flat wings, but sometimes glides with wingtips slightly down.

Sparrowhawks sometimes use an interesting flight mode when hunting birds. They fly directly at a target bird using an undulating flight, apparently to appear similar to a passerine, thus allowing the hawk to close on its prey before the latter tries to evade.

Moult

Annual moult of adults is complete; female begins approximately when clutch is laid and male 2 to 3 weeks later. Some adults suspend moult for a short time while feeding large young. Moult is completed by autumn, except for northernmost birds. Juveniles' moult is also complete, and begins and is completed before that of adults.

Behaviour

Specialize in preying on a wide range of small- to medium-sized birds, which are caught mainly by ambush. They hunt from an inconspicuous perch and fly with a short burst of speed at unwary birds and snatch one as it attempts to fly away. They also hunt by flying slowly and unobtrusively along hedge rows, wood edges, paths, or streams using cover to approach and pounce on unsuspecting birds. Less often they hunt from on high and stoop at birds in open areas. Smaller mammals, carrion, and even worms are also eaten occasionally. Females hunt in more open areas than do males.

Sparrowhawks breed in woodlands, preferring to build their stick nests in coniferous trees. Their courtship flights are typically undulating flights and are done by both sexes. Pairs mutually soar over their territory. Vocalizations are seldom heard outside of nesting area and time.

Status and distribution

Fairly common but unobtrusive breeding residents of forested areas of Europe and north-western Africa but local in the Middle East (Israel). Recently they have moved into some urban forested areas.

Northern populations are migratory, with birds moving south-east into middle and southern Europe for winter. A few winter locally in the Middle East. Populations elsewhere are sedentary.

Fine points

Wing proportions of Eurasian Sparrowhawks and Northern Goshawks are reported to be different but are in fact almost identical. Differences are in wing shape, particularly shape of trailing edge of wing.

Subspecies

Accipiter n. nisus occurs throughout Europe. *A. n. wolterstorffi* is resident on Corsica and Sardinia, *A. n. punicus* is resident in north-west Africa, and *A. n. granti* is resident on Madeira and the Canaries.

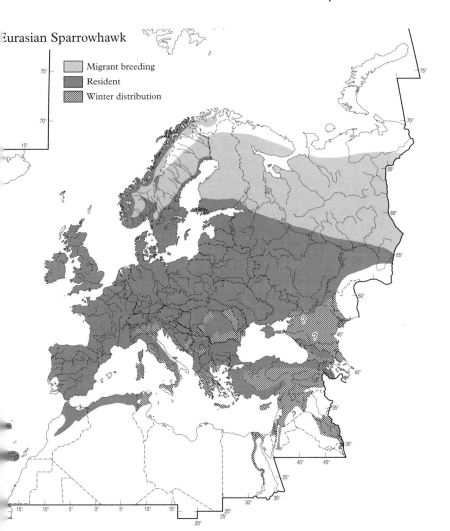

Eurasian Sparrowhawk

Migrant breeding
Resident
Winter distribution

Etymology

'Sparrowhawk' comes from the Old English *spearhafoc* for 'sparrow' and 'hawk', presumably for its habit of preying on sparrows. According to Greek legend, *Nisus*, king of the Megara, was changed into a Sparrowhawk.

Measurements

Length: male 28–32 cm (30 cm)
        female 33–37 cm (35 cm)
Wingspread: male 59–65 cm (62 cm)
           female 69–77 cm (73 cm)
Weight: male 117–160 g (148 g)
        female 220–350 g (260 g)

Northern Goshawk   *Accipiter gentilis*

Plate 23; Photos, p. 322

Description

A powerful, robust hawk, the Western Palearctic's largest *Accipiter* is a denizen of forested areas throughout Europe. Plumages of the three Western Palearctic subspecies differ mainly in colour intensity. Sexes are almost alike in plumage; females are noticeably larger. Juveniles have a distinctly different plumage. *Tapered wings are long* for an accipiter. On perched birds, wingtips extend half-way to tail tip. *Tip of folded tail is wedge-shaped.* Cere is greenish-yellow; robust legs and feet are yellow. The following accounts are for widespread *A. g. gentilis*.

Adult. *Head appears 'hooded' and is slate-grey with wide pearly grey superciliaries* and throat and chin, all of which are somewhat darkly streaked. Back, upperwing, and uppertail coverts, and undersides of flight feathers are dark brownish-grey. Underparts are whitish to pearly grey with bold slate-grey to dark brown barring, often with narrow dark shaft streaks. Females average browner on upperparts and more heavily marked on underparts. Underwings show pearly grey coverts with slate-grey barring and greyish flight feathers with more or less distinct dark banding that is less bold than that of juvenile and other accipiters. Upperside of tail usually appears unmarked brownish-grey but sometimes shows dark bands (younger adults?); underside appears paler with three or four wide dark bands. Wide white band is noticeable on tip of fresh tail feathers. Unmarked undertail coverts are white and fluffy, but are occasionally banded. Iris is orange-red in older males and orange-yellow in females and younger males.

Second winter birds are essentially like adults but have wider barring on underparts, more brownish-grey upperparts, and a few retained juvenile upperwing coverts or flight or tail feathers. Iris is orange-yellow to bright yellowish.

Juvenile. Head is brown except for wide creamy superciliaries and creamy throat with narrow buffy streaking on crown and cheeks and narrow dark brown streaking on throat. Darker malar stripes are noticeable. Dark brown back and upperwing and uppertail coverts appear paler because of wide cinnamon-rufous and buffy feather edges. *Two short pale bars are usually visible on upperwing coverts on each wing of flying hawks.* Uppersides of flight feathers are dark brown, with faint darker barring. Rufous-buff (fresh) to creamy (faded) underparts have medium to heavy brown streaking that becomes finer on belly and narrow dark shaft streaks on leg feathers and undertail coverts. Buffy underwings show dark brown bands on flight feathers and dark brown streaks on coverts. *Tail is composed of wavy, equal-width dark and light brown bands, with thin white 'highlights' between many bands and a wide white terminal band that emphasizes the wedge shape of tail tip.* Underside of folded tail (outer feathers) is pale with narrower dark brown bands. Iris is light brown to yellow.

Note: Goshawks vary in size and colour intensity from larger and paler in north and east to smaller and darker in south and west; adults vary from browner in west to greyer in east. Subspecies *buteoides* is paler overall and is less heavily barred than is nominate gentilis; *arrigonii* is overall darker.

Unusual plumages
Partial albino and dilute plumages have been reported.

Similar species
Eurasian Sparrowhawks (Plate 23) can appear similar in shape and plumage but are smaller and less robust; they have less tapered wings, more stick-like legs, more boldly marked underwings, and square-tipped tails that have at most a narrow pale band on tip; they lack hooded head of adult Northern Goshawks. Juvenile Eurasian Sparrowhawks are always barred on underparts and underwing coverts, whereas juvenile Northern Goshawks are streaked.
Gyrfalcons (Plate 45) can appear similar to adult Northern Goshawks. See under that species for distinctions.

Flight

Powered flight is strong, with a series of powerful, stiff wingbeats followed by a short glide; then series is repeated. Soars frequently with wings held level; glides with wrists forward and wings level.

Moult

Annual moult of adults is usually complete, beginning in spring and completed by end of summer. Not all rectrices or secondaries are moulted every year. Post-juvenile moult is not complete and begins later than moult of adults. Often juvenile flight or tail feathers or upperwing coverts are retained.

Behaviour

Aggressive, agile predators that use many hunting techniques to capture medium to large birds and mammals. Prey is taken on ground and from trees, or snatched in flight. They hunt from both conspicuous and inconspicuous perches, from ground-hugging flight, and from high on soar. Prey is usually approached directly in rapid flight, but some, particularly adults, approach indirectly, taking advantage of available cover to get close enough to surprise prey. If prey is missed on first attempt, it is often tail-chased. During winter, Northern Goshawks sometimes eat carrion.

They build nests high in tall trees, usually in extensive forests. Courtship displays include undulating flight, sky-dancing, and slow flapping. Displays also include displaying fluffed out white undertail coverts. Usually silent, they vocalize often while breeding.

They are quite secretive and are not easily seen, as they spend much time perched inconspicuously.

Status and distribution

An uncommon but widespread resident of open woodlands over most of Europe but rare and local in England, Scotland, and Turkey. They also occur in a small area of north-western Morocco.

They breed in a variety of wooded habitats from northern coniferous forests to southern broad-leafed ones and from sea-level to tree-limit in mountains.

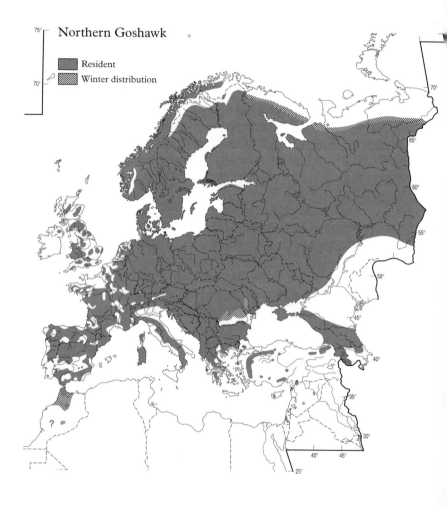

Northern Goshawk

Resident
Winter distribution

Only some northernmost adults leave their breeding areas during winter, depending on availability of cyclic prey. Juveniles disperse widely, most northern ones moving south for first winter.

Fine points

Buffy tips of juveniles' auricular feathers form a faint but noticeable owl-like facial disk.

Subspecies

A. g. gentilis is widespread throughout the Western Palearctic. *A. g. buteoides* breeds from northern Sweden and Finland eastward across northern Russia. *A. g. arrigonii* is permanent resident on Corsica and Sardinia.

Etymology

'Goshawk' comes from the Anglo-Saxon *gos* for goose and *havoc* for hawk. *gentilis* is Latin for 'noble'. It was named during an era when only the nobility could fly this species in falconry.

Measurements

Length: 46–62 cm (54 cm)
Wingspread: 96–115 cm (105 cm)
Weight: 615–1450 g (1050 g)

BUZZARDS—GENUS *BUTEO*

Three species of buzzards occur in the Western Palearctic. All are medium-sized, with robust bodies, relatively small beaks, long, broad wings, and short to medium length tails. All three soar and hover regularly but are not usually dashing in flight and spend most of their time perched.

Two species show a wide variation in plumages; one of these, *Buteo buteo*, has two recognizable subspecies, Common Buzzard and Steppe Buzzard. Many individuals of all species show a characteristic dark spot in the centre of the nape.

Tails of juvenile buzzards average longer than those of adults, as in many raptors. However, unlike other raptors, wings of juveniles are, on average, narrower than those of adults.

'Buzzard' comes from the same Latin root through Old French and Old English; *Buteo* is Latin for 'a kind of hawk or falcon'.

Common Buzzard *Buteo b. buteo*
(Eurasian Buzzard)

Plates 24–27; Photos, p. 324

Description

A common, widespread resident throughout Europe, but intergrades with Steppe Buzzard in eastern Europe and Finland. Sexes are alike in plumage; females are noticeably larger. Plumages of juveniles differ from those of adults. Wingtips reach tail tip on perched buzzards. Iris of adults is dark brown; that of juveniles is pale brown. Cere is yellow and legs are pale orangish-yellow. Plumages are dimorphic, with normal-morph birds common and occurring throughout western Europe and white-morph individuals occurring locally, mainly in southern Sweden and northern Germany and Poland but also in Britain. Latter differ in being quite variable in plumage, with a wide range of variation from normal-morph birds to some almost completely white individuals.

Normal-morph adult. Head is usually uniformly medium brown to dark blackish-brown, but paler birds may show whitish throat streaking. Back, upperwing and uppertail coverts, and flight feathers are medium brown to dark blackish-brown; latter may show even darker banding when viewed close. Tail above appears brown with a greyish cast and shows numerous narrow darker bands and a wide dark subterminal band. Breast is usually uniformly medium brown to dark blackish-brown, but on paler birds may show some whitish streaking. Belly is whitish with a variable amount of dark barring, heavier on flanks; most birds show an unbarred area between breast and belly that appears as a pale 'U' on undersides. Leg feathers are uniformly medium brown to blackish-brown. Whitish undertail coverts show dark barring. Underwing is two-toned: coverts vary from uniformly dark blackish-brown to heavily barred dark brown, with primary coverts (carpal patches) usually appearing a darker blackish-brown, and greyish flight feathers show a wide dark subterminal band and several narrow dark bands on secondaries and inner primaries. Outer four primaries are unbarred below and appear whitish. Tail below appears whitish with a wide dark subterminal band and numerous other narrow bands.

Normal-morph juvenile is like adult but with more pale streaking on breast; markings on belly are dark spots or streaks. Underwing and tail patterns are different: underwing lacks wide dark subterminal band and has instead a narrower dusky subterminal band; brown tail has many equal-width dark bands, with subterminal same width or only slightly wider than others. Tail bands are usually wider than the narrow bands on tails of adults.

White-morph adult. Plumages are quite variable. Head varies from brown to completely white. Back, upperwing and uppertail coverts, are usually brown but can have few to many partially or completely white feathers or little to extensive buffy and rufous feather edges. Breast is often uniform brown forming a noticeable bib. White belly is usually unmarked, but may show a dark patch on flanks. Leg feathers vary from completely white to white with rufous-brown barring. Undertail coverts are usually white and unmarked. Underwing is whitish with a dark 'comma' or carpal patch and a wide dark subterminal band on trailing edge of wings. Tail pattern is same as normal morph, but base colour is white, light grey, or rufous.

White-morph juvenile. Plumages are similar to those of adult and, like them, also are quite variable. Differs from adult in having a narrower, more dusky band on trailing edge of wing, tail with equal-width tail bands, and pale brown iris.

Unusual plumages
Albinism and partial albinism have been reported. Two dilute-plumage birds have been prepared as museum mounts, one in England and one in Italy.

Similar species
Eurasian Honey Buzzards (Plates 6, 7) can appear somewhat similar to Common Buzzards. See account for that species for distinctions.

Rough-legged Buzzards (Plates 24–26) can appear similar to paler Common Buzzards. See under that species for distinctions.

Short-toed Snake Eagles (Plate 5) can appear similar to paler Common Buzzards but are much larger, with relatively shorter, more square-cut tails, and lack dark carpal patches or commas on underwings.

Flight

Powered flight is with fairly quick, rather shallow wingbeats of somewhat stiff wings. Soars with wings held in a medium dihedral; glides with wings held level or with wingtips lowered a bit. Like most buzzards, they hover regularly.

Moult

Annual moult of adults is complete during spring to autumn. Juveniles of this race also go through complete moult.

Behaviour

Typical buzzards that as such are general feeders, preying on small mammals, birds, reptiles, amphibians, insects, and even earthworms. In winter they also eat carrion. Hunting is usually from an elevated perch such as tree top, power pole, or fence post, but they also hunt from soar or glide and frequently hover. They are comfortable standing or walking on the ground, especially when catching insects.

Their stick nests are usually placed in trees, but cliff ledges and rocky crags are also used. Display flights are typical and included circling high over territory, undulating flight, and deep wingbeat flight. They are quite vocal on territory, especially when breeding.

Status and distribution

Widespread and fairly common throughout most of Europe, except for Iceland, Ireland, and northern Scandinavia and Russia, and across northern Turkey and the Caucasus. Northern and western buzzards are migratory, with Scandinavian buzzards moving south and west into Iberia and north-western Africa, and Finnish and Russian ones moving south into Turkey, the Middle East, and Africa.

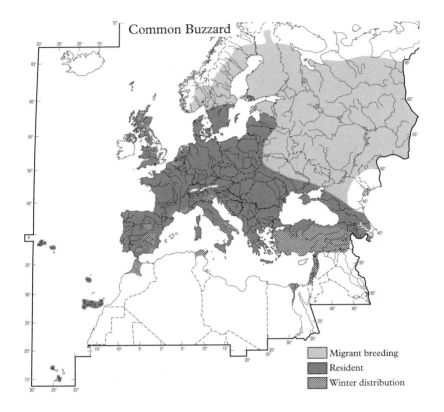

Common Buzzard

Migrant breeding
Resident
Winter distribution

Counts of many buzzards have been recorded in autumn in eastern Turkey as they pass the eastern end of the Black Sea.

Populations in some areas, particulary south England, are much reduced due to human persecution, primarily by shooting and poisoning.

Fine points

Some whitish adults strongly resemble some adult male Rough-legged Buzzards as both can show dark breast and white, relatively unmarked bellies, but they lack the dark eye-lines, bright orange cere, feathered tarsi, and white forehead spot of the Rough-leg.

Subspecies

Two widespread races, *B. b. buteo* and *B. b. vulpinus* occur in the Western Palearctic, plus four insular races: *B. b. rothschildii* in the Azores, *B. b. insularum* in the Canaries, *B. b. bannermani* in the Cape Verdes, and *B. b. arrigonii* on Corsica and Sardinia. See James (1984) for a discussion of variation in plumages of these insular forms.

Etymology

Buzzard comes from the Old French *busard*.

Measurements

Length: 46–53 cm (49 cm)
Wingspread: 115–137 cm (127 cm)
Weight: 600–1350 g (890 g)

Steppe Buzzard *Buteo b. vulpinus* (Eurasian Buzzard)

Plates 24–7; Photos, p. 326

Description

A common, widespread resident of Asia, intergrading in eastern Europe and Finland with the Common Buzzard. They are noticeably smaller than Common Buzzards; intergrades between races occur commonly in the extensive zone of contact. Sexes are alike in plumage; females are noticeably larger. Plumages of juveniles differ from those of adults. Wingtips reach tail tip on perched buzzards. Iris of adults is dark brown; that of juveniles is pale brown. Cere is yellow and legs are pale orangish-yellow. Plumages are polymorphic: dark, rufous, and grey-brown morphs, with intergradation among them. Most lack square black carpal patches on underwings, usually having only a dark comma.

Rufous morph

Rufous morph is most common and is quite variable.

Rufous-morph adult. Head is rufous-brown to rufous. Back and upperwing and uppertail coverts are brown with variably sized rufous-buffy feather edges. Uppersides of primaries are dark grey-brown. Upperside of tail is bright rufous, often with many narrow dark bands and a wide subterminal band, but some tails are unbanded. Underparts vary from uniformly pale to dark rufous to pale to dark rufous on breast and pale to dark rufous barring on white belly and undertail coverts, in which case they usually show a pale 'U' between breast and belly. Leg feathers are usually uniformly pale to dark rufous. Underwings show pale to dark rufous on lesser and median secondary coverts. Greater secondary coverts are paler and show dark brown barring. Primary coverts are rufous except for tips of greater coverts, which are usually blackish and form a dark comma on each underwing. Whitish flight feathers have a wide dark terminal band; secondaries and inner primaries also show three rows of narrow dark banding. Undertail pattern is same as that of uppertail but in colour appears a paler rufous.

Rufous-morph juvenile. Overall appear paler and less rufous that do adults of this morph. Head is pale to medium brown with whitish throat. Back and upperwing and uppertail coverts are brown and usually lack any rufous feather edges. Secondaries above are same colour, but primaries are paler, forming a panel on each upperwing. Underparts are streaked brown and white, with an unmarked area between breast and belly forming a pale 'U'. Leg feathers are usually dark brown. Whitish undertail coverts are variably streaked. Tail is light brown with numerous equal-width dark bands. Underwing is pale with some dark streaking on coverts; black tips of greater primary coverts form a dark comma on each underwing, three narrow dark bands through secondaries, and a dusky band on trailing edge of wing that is narrower than that of adult. Iris is coloured straw to pale brown.

Grey-brown morph

Grey-brown morph is fairly common and in pattern is similar to Common Buzzard.

Grey-brown adult. Pattern is like that of Common Buzzard adult with dark blackish-brown coloration replaced by paler grey-brown. Carpal patch usually is not solidly black, but with only a dark comma, tips of greater primary coverts. Tail is usually a mixture of rufous and grey with the adult Buzzard pattern of wide dark subterminal band and numerous narrow dark bands.

Grey-brown juvenile. Pattern is like normal-morph Common Buzzard with dark blackish-brown replaced by pale to medium brown. Carpal patch usually is not solidly black, but with only a dark comma, tips of greater primary coverts. Brownish tail is with typical juvenile Buzzard pattern of equal-width dark bands.

Dark morph

Dark morph is the least common.

Dark-morph adult. Head, back, underparts, all coverts, and leg feathers are uniformly dark blackish-brown to jet black, except for greater secondary coverts on underwing, which are greyish with some whitish barring. Flight feathers are dark brown on upperside but whitish on underside, with wide black terminal band and three or four rows of narrow black bands on secondaries and inner primaries. Greyish tail has wide black subterminal band and numerous narrow black bands.

Dark-morph juvenile is similar to adult, but overall coloration is usually paler and more brownish, lacking in blackish tones. Some birds have some whitish streaking, particularly on breast and underwing coverts. Like all juvenile Buzzards, they have tails with equal-width dark bands, narrower dusky band on trailing edge of wings, and pale brown iris colour.

Unusual plumages
Albinism and partial albinism have been reported.

Similar species
Long-legged Buzzards (Plates 24–27) are often very difficult to separate from rufous-morph and dark-morph Steppe Buzzards. Best field marks of Long-legs are larger size, relatively longer, eagle-like wings, longer neck and head projection, and wings held in more noticeable dihedral when soaring. Rufous-morph Long-legs show black carpal patches in all but palest plumages, a feature lacking in most rufous Steppe Buzzards (but may be present on some intergrades with *B. b. buteo*), and have darker markings across belly and flanks. Dark-morph birds of both forms are most difficult to distinguish; best field marks are mentioned above. However, secondaries of dark Long-legs tend to be dark with irregular whitish markings, whereas those of dark Steppes tend to be whitish with regular dark banding. Tail pattern of juvenile dark Long-legs is most often dark with narrow whitish bands; that of juvenile dark Steppes is brown with even-width darker brown bands.

Flight
Powered flight is with fairly quick, rather shallow wingbeats of somewhat stiff wings. Soars with wings held in a medium dihedral; glides with wings held level or with wingtips lowered a bit. Like most buzzards, they hover regularly.

Moult

Annual moult of adults is begun on breeding grounds, but is suspended during long migratory flight and continued on winter quarters in Africa. Moult of all feathers is probably not completed annually. Juveniles of this race do not begin moult on winter quarters, but do so during return migration in spring. Like adults, they do not complete moult during summer and suspend it during migration and continue on winter quarters. First plumage adults retain a few juvenile secondaries into their third summer.

Behaviour

Typical buzzards that are general feeders, preying on small mammals, birds, reptiles, amphibians, insects, and even earthworms. In winter they also eat carrion. Hunting is usually from an elevated perch such as tree top, power pole, or fence post, but they also hunt from soar or glide and frequently hover. They are comfortable standing or walking on the ground, especially when catching insects.

Their stick nests are usually placed in trees, but cliff ledges and rocky crags are also used. Display flights are typical and include circling high over territory, undulating flight, and deep wingbeat flight. They are quite vocal on territory, especially when breeding.

Status and distribution

Widespread and fairly common throughout very eastern Europe and completely migratory. Large numbers of Steppe Buzzards from eastern Europe and western Asia pass through eastern Turkey and the Middle East, especially Israel, Suez, and Sinai, in autumn and spring.

Fine points

Some authorities tout the presence of pale primary patches on upperwings as a field mark of Long-legged Buzzard not Steppe Buzzard, but there is overlap of this character between these buzzards. Additionally, some authorities report longer tail of Long-legged Buzzard compared to that of Steppe Buzzard, but measurements of live birds show that tails of both are proportional to overall lengths.

Subspecies

Two widespread races, *B. b. buteo* and *B. b. vulpinus*, occur in the Western Palearctic, plus four insular races: *B. b. rothschildii* in the Azores; *B. b. insularum* in the Canaries; *B. b. bannermani* in the Cape Verdes; and *B. b. arrigonii* on Corsica and Sardinia. See James (1984) for a discussion of variation in plumages of these insular forms.

Etymology

Buzzard comes from the Old French *busard*. *vulpinus* is Latin for 'fox-like', for tawny plumage.

Measurements

Length: 40–47 cm (44 cm)
Wingspread: 100–125 cm (114 cm)
Weight: 400–950 g (715 g)

Long-legged Buzzard   *Buteo rufinus*

Plates 24–27; Photos, p. 328

Description

A large, long-winged buzzard, widespread resident of south-eastern Europe and north-western Africa, but local in the Middle East. It is polymorphic, having light, rufous, and dark morphs, but with much intergrading among morphs. A pale nape patch with a black spot in its centre occurs on almost all birds; pale-headed birds show only the dark spot. All but the palest birds show a square black carpal patch on each underwing. The two races cannot be distinguished in the field. Sexes are alike in plumage; females are noticeably larger. Plumages of juveniles differ from those of adults. Wingtips reach tail tip on perched birds.

Light-morph adult. Head is usually quite pale, varying from white to pale rufous and, except on extremely pale birds, shows narrow dark eye-line and malar stripes. Back and lesser and median secondary upperwing covert feathers have dark centres and wide rufous edges; in total they appear quite rufous overall and contrast with darker greyish-brown uppersides of flight feathers and dark brown primary and greater secondary coverts. Many birds show a variably sized pale patch on upper primaries. Lower back and lesser uppertail coverts are dark brown. Greater uppertail coverts and upper side of tail are rufous; latter is usually unbanded. Pale birds usually show rufous tails with white bases. Breast varies from white or creamy to pale rufous, usually with some short, narrow dark streaking. Belly, flanks, and leg feathers are dark rufous to dark brown and contrast with pale breast. Undertail coverts and undertail appear whitish to pale rufous. Underwings show square black carpal patches, creamy to pale rufous coverts often with short dark streaking, and whitish flight feathers whose black terminal bands form a wide dark border to trailing edge of wing. Paler birds often have dark on carpal area restricted to greater coverts; this then appears as a dark comma. Adults have dark brown eyes.

Light-morph juvenile. Head is creamy with a variable amount of rufous or brown streaking; as a result, it can appear creamy, light brown, or pale rufous. Back and upperwing coverts are brown, usually with little or no rufous feather edging, and do not contrast strongly with dark uppersides of flight feathers. Almost all birds show pale patches on upper primaries. Greater uppertail coverts vary from mostly rufous to brown with rufous tips. Upper tail is greyish brown, usually with four or five equal-width narrow dark brown bands on lower half; darker birds have bands on entire tail. Paler birds often show a rufous cast to upper tail. Creamy underparts are streaked with dark brown, with a U-shaped unstreaked area between breast and belly noticeable on all but palest birds. Flanks and leg feathers are more heavily marked, varying from rufous to uniform dark brown. Creamy undertail coverts may show some dark markings. Underside of tail appears creamy with narrow dark bands. Underwings are like those of adults except that band on trailing edge is narrower and is dusky, not black. Juveniles have very pale brown eyes.

Dark-morph adult. Head, body, all coverts, and leg feathers are jet black to blackish-brown. Many birds show a pale nape patch; some show pale or rufous streaking on centre of breast. Greyish panel on primaries is noticeable on upperwing. Flight feathers below are whitish-grey with a wide black terminal band and three or four narrow black bands that span across all feathers. Tail is whitish-grey with a wide black terminal band and, usually but not always, four or five narrow black bands across all feathers. Adults have dark brown eyes.

Dark-morph juvenile is like dark-morph adult except that overall colour is more brownish and less blackish and underwing and tail patterns differ. Underwing pattern is similar to that of adult, but markings are more brownish and are heavier on secondaries. Tail is dark brown with four narrow whitish bands across each feather. Juveniles have very pale brown eyes.

Rufous-morph adult. Head is dark rufous. Back and upperwing coverts are brown with some rufous-buffy feather edges and usually contrast somewhat with dark uppersides of flight feathers. Pale primary panel is same as in dark morph. Dark brown greater uppertail coverts have wide rufous tips. Rufous or grey tail has same pattern of dark banding as that of dark morph. Underwing coverts are rufous, carpal patches are black, and flight feathers are like those of dark morph. Breast is rufous and belly, flanks, and leg feathers are dark rufous to dark rufous-brown. Undertail coverts are rufous to dark brown. Adults have dark brown eyes.

Rufous-morph juvenile is like the rufous-morph adult, but some birds have dark or pale breast streaking, others lack rufous edges on back and upperwing coverts, and all have juvenile tail pattern. Underwing pattern is much like that of dark-morph juvenile, whereas tail pattern is like that of light-morph juvenile.

Unusual plumages

No unusual plumages have been reported.

Similar species

Common Buzzard (Plates 24–27) is easily separated from the larger Long-legged Buzzards by the lack of rufous in plumage. Also, it has no dark or rufous morphs.

Steppe Buzzard (Plates 24–27) is very similar in most plumages to Long-legged Buzzards. See description under account for that species for distinctions.

Rough-legged Buzzard (Plates 24–26) can appear somewhat similar to Long-legged Buzzards, but these species seldom occur together at the same location. Rough-legs lack rufous in plumage and do not have a dark morph.

Bonelli's Eagle juveniles (Plate 36) can appear similar in colour to juvenile Long-leggeds but lack square black carpal patches and dusky band on trailing edge of underwings.

Flight

Powered flight is with slow, deliberate wingbeats of somewhat flexible wings. Soars with wings held in a medium to strong dihedral, and glides with wings held in a modified dihedral. Like most buzzards, Long-legged Buzzards hover regularly.

Moult

Annual moult has not been well studied. Apparently that of adults is complete, but on migrants it is suspended during migration and completed on wintering quarters. Juveniles begin moult to adult plumage in late spring and usually complete this by autumn.

Behaviour

Prey mainly on small mammals, reptiles, and large insects but also on frogs and toads, snakes, and birds. In winter they have been reported to eat carrion. They hunt from the air in a glide or soar or by hovering but, more often, they hunt from exposed perches such as rock outcrops, small trees, poles, and hay bales. Sometimes they walk around on the ground to capture insects or wait at mammal burrows.

They build their stick nests on small trees, crags, or cliff ledges, or, when none of these are available, on the ground.

Their display flights are typical of buzzards, with high circling and undulating flights reported, often accompanied by vocalizations, which are little known and apparently similar to those of Common Buzzard.

Status and distribution

Fairly common residents of arid areas of south-eastern Europe, north-western Africa, and the Middle East, including arid steppe, semi-deserts, and low mountains. They are summer residents and migrate from Bulgaria, Macedonia, Greece, and Russia for winter, travelling into northern Africa, some going south of the Sahara. They are resident in Turkey and locally in Syria, Lebanon, Jordan, and Israel and are resident also in Morocco, northern Algeria, Tunisia, and locally in Libya and Egypt.

Migrants pass through Israel and Suez, Egypt in some numbers in both autumn and spring.

Long-legged Buzzard

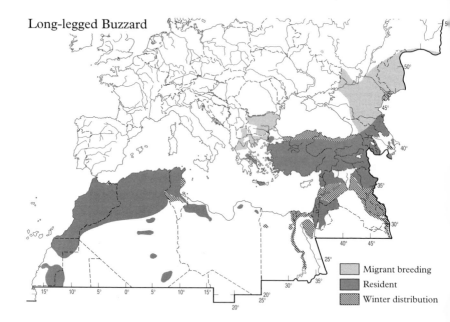

Migrant breeding
Resident
Winter distribution

Fine points

Some authorities tout the presence of pale primary patches on upperwings as a field mark of Long-legged Buzzard not Steppe Buzzard, but there is overlap of this character between them. Additionally, some authorities report longer tail of Long-legged Buzzard compared to that of Steppe Buzzard, but measurements of live birds show that tails of both are proportional to overall lengths.

Subspecies

Both subspecies occur in the Western Palearctic. Nominate *rufinus* in Europe and the Middle East and *cirtensis* in north-western Africa. The latter differs from the nominate mainly by its smaller size and lack of dark and rufous morphs.

Etymology

rufinus is Latin for 'reddish'.

Measurements (nominate)

Length: 51–57 cm (54 cm)
Wingspread: 136–159 cm (148 cm)
Weight: 850–1425 g (1095 g)

Rough-legged Buzzard   *Buteo lagopus*

Plates 24–26; Photos, p. 364

Description

A summer inhabitant of arctic tundra of northernmost Europe and a winter resident in its middle latitudes. *Legs are feathered down to toes.* Plumage characters distinguish sexes of adults; however, some buzzards have one or more plumage characters of other sex. Females are larger than are males. Juvenile plumage is different from those of adults. Most birds have a square black carpal patch on otherwise whitish underwing. On perched birds, wingtips reach tail tip on juveniles but project beyond tip on adults.

Adult male type. Head appears somewhat dark with blackish face and eye-line and shows a variable amount of buffy streaking on crown and face; pale nape patch is noticeable. Back and upperwing coverts are dark blackish-brown with much whitish, greyish, and rufous mottling. Uppersides of flight feathers are grey; adults lack white primary patches on upperwings, showing at most three narrow white streaks (inner webs of outer three primaries). Uppertail coverts are white with large black spots. White tail has a wide dark subterminal band and one or more usually incomplete narrower dark bands, with dusky grey or rufous-grey colour between bands. Dark blackish-brown breast has a variable amount of buffy streaking and usually appears uniform dark as a bib. Whitish belly and flanks are barred with dark blackish-brown, but bellies on some birds are unmarked. Adults show a pale, unmarked 'U' between breast and belly. Underparts of males usually have breast more heavily marked than belly. Whitish leg feathers are heavily barred with dark blackish-brown. Whitish undertail coverts are lightly marked. Whitish underwings of adults show dark markings on coverts, black carpal patch with whitish mottling, and some darker banding on flight feathers with a noticeable wide dark band on trailing edge. Adult iris is dark brown. Cere and legs are orange.

Adult female type is similar to adult male type but can be distinguished by pattern and colour of head, back, underparts, and tail.

Head usually appears paler, with brownish streaking and narrower dark face. Back is dark brown with whitish and rufous mottling but lacking grey tones. Creamy breast has dark streaking, and belly and flanks are uniform dark blackish-brown forming a belly band; some birds show a clear area in mid-belly. Females are usually more heavily marked on belly than on breast. Creamy legs feathers show dark blackish-brown spotting or barring. Undertail coverts are lightly marked. White tail above shows wide dusky tip with narrower darker subterminal band superimposed and may show one or two other narrow dark bands; on underside tail appears pale with dark subterminal band. Dark carpal patches on underwings usually have little or no whitish mottling. Cere and legs are orange-yellow.

Juvenile appears similar to adult female, but iris is pale brown, breast and underwing coverts are usually less heavily marked, tail has only dusky band, belly band lacks blackish tone, and trailing edge of underwing shows narrower dusky band. Creamy leg feathers show only narrow dark streaking. Upperwings show large whitish patches on primaries. Cere and legs are yellow.

Unusual plumages

No unusual plumages have been reported.

Similar species

Common Buzzards (Plates 24–27), particularly whitish birds, can appear similar, but are smaller and shorter winged and lack dark eyelines. Common Buzzards with white in tail do not have dark belly bands; they also fly with stiffer wings and glide with wings level.

Long-legged Buzzards (Plates 24–27) can appear similar. See under that species for distinctions.

Golden Eagle juveniles (Plate 35) (and some older immatures) can also show whitish tail with dark tip and white primary patches on upperwings but are much larger and more aquiline and have uniformly dark underparts.

Flight

Powered flight is with slow, rather shallow wingbeats on flexible wings. Soars with wings in a medium dihedral; glides with them in a modified dihedral. Hovers frequently; sometimes with deep wingbeats, sometimes with fluttering wings.

Moult

Annual moult of adults is complete except that all flight feathers are not replaced every year. Moult begins in May or June and is usually completed by autumn migration but may be completed in winter area. Post-juvenile moult is also complete except for some flight feathers and begins earlier in spring than does that of adult.

Behaviour

Specialists that prey on small mammals, particularly rodents, which they hunt from perches and by hovering and occasionally by standing on ground. They shift to preying on birds in times of mammal scarcity but are usually unable to raise young on this prey. They occasionally eat carrion in winter.

Rough-legs are usually solitary but will concentrate to breed or in winter in areas of locally abundant prey, even forming communal night roosts in winter.

Nests are usually constructed on a cliff, crag, or small pinnacle, but also in small trees and bushes, and, in absence of these, on the ground. Pair perform display flights during the short courtship period, including undulating flight. Vocalizations are similar to those of Common Buzzard but are more drawn out. Routh-legs are usually silent when not breeding.

Status and distribution

Locally common summer breeders in high latitudes of arctic and subarctic treeless tundra from western Norway across northern parts of Sweden, Finland, and Russia into Asia, concentrating where small mammal populations are peaked. In years of abundant prey (rodents) breeding extends into open taiga areas. Entire population moves south to south-east to middle latitudes of Europe for winter,

Rough-legged Buzzard

Migrant breeding
Winter distribution

ranging from coastal England and Scotland (rarely) south and east from northern France and southern Scandinavia across to the former Yugoslavia and the Black Sea. They are fairly common in open treeless areas.

Vagrants have been recorded in Iceland, Ireland, and the Faroes. Stragglers regularly reach most countries bordering the Mediterranean Sea, including Cyprus and Malta.

Fine points

Buteo l. sanctijohannis, the North American subspecies, has a dark colour morph with distinct plumages for each age/sex class.

Subspecies

Nominate *Buteo l. lagopus* occurs in the Western Palearctic.

Etymology

Named 'rough-legged' for its completely feathered tarsi. *lagopus* is from the Greek *Lagos* meaning 'hare' and *pous* for 'foot'.

Measurements

Length: 46–59 cm (53 cm)
Wingspread: 122–143 cm (134 cm)
Weight: 745–1600 g (1200 g)

BOOTED EAGLES—GENERA *AQUILA* AND *HIERAAETUS*

Seven species of eagles in the genus *Aquila* and two eagles in the genus *Hieraaetus* occur regularly in the Western Palearctic. All have legs feathered to the toes and are notoriously difficult to identify correctly in the field. Juveniles are usually more distinctly marked and easier to recognize. All are often seen soaring and, except for the Bonelli's and Spanish Imperial Eagles, are somewhat migratory. An eighth *Aquila* species, the Verreaux's or Black Eagle, occurs barely in the Western Palearctic locally in Jordan and the Sinai.

Many individuals show white bases on the uppersides of the primaries that form a small whitish patch, the so-called *'Aquila'* patch. Some individuals of Steppe, Tawny, Lesser Spotted, and Greater Spotted Eagles show a small white spot on the lower back.

The five largest, Golden, Eastern Imperial, Spanish Imperial, Steppe, and Tawny Eagles, reach adult plumage at 4 or 5 years of age. The Bonelli's and both spotted eagles reach adult plumages after 3 or 4 years.

Tawny and Booted Eagles differ from the others in being polymorphic, with much plumage variation among same-age birds.

The two spotted eagles differ by having round, not slotted, nostrils and having short, dense feathering on the tarsus, which gives their legs a 'stove-pipe' look.

'Eagle' comes from the Middle English *egle* and the Old French *egle* or *aigle*, which came from the Latin *aquila*, for 'eagle'.

Lesser Spotted Eagle   *Aquila pomarina*

Plates 28, 30; Photos, p. 330

Description

A medium-sized dark eagle that is a summer breeding resident in the north central to south-eastern Western Palearctic. In all plumages they appear similar to but usually paler than Greater Spotted Eagles. Both spotted eagles have rounded nostrils (oval or elliptical on other eagles) and short, tightly feathered tarsi (longer and looser in other eagles). The underwings of Lesser Spotted Eagles almost always show secondary coverts paler than the flight feathers; Greater Spotted Eagles usually show coverts darker or the same colour as the flight feathers but can show paler coverts. When perched, the heads of Lessers appear noticeably smaller than those of Greaters; their shorter nape feathers give their heads a more rounded appearance, as well. Compared to other *Aquila* eagles, the two spotted eagles are noticeably smaller, and in flight show relatively shorter head projection, wings, and tails. Almost all Lessers show the white *Aquila* patch at base of primaries, and a white carpal comma or two is usually visible on underwings. Inner primaries are somewhat paler on most eagles. Wingtips reach or almost reach tail tip on perched eagles. Tail is dark brown with a narrow pale terminal band. Beak is two-toned, dark at tip and horn-coloured at base. Cere, gape, and toes are yellow to pale yellow. A pale colour morph has been reported, but, as it has not been adequately described and specimens of it do not exist, it should be considered hypothetical.

    Adult. Overall medium brown (but occasionally darker brown), except for paler head, paler brown secondary wing coverts, white tips on greater uppertail coverts, and pale tips of tail feathers. Pale upperwing coverts contrast with darker back and flight feathers. Bases of inner primaries are pale on uppersides and form the *Aquila* patch. Undersides of flight feathers are dark greyish-brown, usually with faint pale barring and appear darker than the underwing coverts. *Undertail coverts have narrow buffy tips.* Iris is golden yellow.

Juvenile. Head is medium brown, with paler face and rufous patch on nape and often some rufous tips on nape feathers. Medium brown backs and upperwing coverts usually have small buffy streaks or spots on tips; two rows of larger whitish spots are noticeable on tips of median and greater coverts (the name 'spotted' is derived from these); spots are usually not as large nor as numerous as those of Greater Spotted Eagles, but overlap is possible (see Plate 30). Greater uppertail coverts are white and form a pale 'U' at base of upper tail; other uppertail coverts and rump are medium brown. Underparts and upper leg feathers are medium brown. Narrow buffy streaking on underparts is heaviest on the breast, sometimes forming a large pale breast patch. Undertail coverts are whitish. Underwings show contrast between paler brown lesser and median coverts and darker greyish-brown greater coverts and flight feathers, the latter covered with narrow blackish barring. Tarsi feathers are usually buffy. Iris is dark brown.

Second winter eagles appear almost identical to juveniles in that the new feathers are identical to the replaced ones, except for being darker and unworn. Rufous nape patch still noticeable. New secondaries and tail feathers have wider white tips and differ from the retained ones, whose tips are often worn, resulting often in ragged uneven trailing edge of wings. Pale spots on upperwing coverts are also often worn and now appear, if at all, as short narrow whitish streaks. Iris is brown.

Third winter eagles are transitional, with a mixture of new, adult-like and old immature feathers. New nape feathers lack pale tips, and new feathers on underparts lack buffy streaking. New upperwing coverts are brown lacking white spots; old ones show at most small white spots or short streaks. New secondaries are longer and darker, usually with less distinct banding. New greater uppertail coverts are white; white 'U' at base of tail is still noticeable. New undertail coverts are medium brown with pale edges. Iris is becoming more yellow.

Fourth winter eagles are essentially adults, but some may retain a few immature feathers. New uppertail coverts are medium brown with whitish tips. New undertail coverts are brown with buffy tips.

Unusual plumages

Hybrids with Greater Spotted Eagles have been reported. If true, then the offspring would show characters of both. No other unusual plumages have been reported.

Similar species

Greater Spotted Eagle (Plates 29, 30) is similar to Lesser Spotted Eagle in all plumages. See under that species for distinctions.

Steppe Eagle (Plates 31, 32) older immatures can appear similar in flight to Lesser Spotted Eagles but show a wide pale band on under-wings and have relatively longer wings, tails, and head projection.

Flight

Powered flight is with rather buzzard-like wingbeats of somewhat cupped wings. Soars with wings held level or with wingtips a bit depressed. Glides with wings level to wrist and wingtips somewhat depressed.

Moult

They apparently undergo an incomplete annual moult, in that not all flight feathers (perhaps not all body feathers, as well) are replaced every year, but detailed studies are lacking. Moult is most likely commenced in early spring and suspended for the migration in autumn; however, breeding adults may suspend their moult for a period in summer. Moult is completed on the winter grounds in Africa.

Behaviour

Opportunistic predators, taking a wide range of prey, primarily small mammals, but also amphibians and smaller numbers of birds, as well as a small number of reptiles and insects. In winter in Africa, they regularly feed on emerging termites and nesting Queleas.

Hunting is from perches, on the wing, and by walking on the ground. They are usually somewhat sluggish and take easily captured prey, usually capturing it on the ground.

Flight displays include undulating flight and mutual soaring. Typical raptor stick nests are built on trees and, although two eggs are laid, only one eaglet is raised, as the older chick kills the younger in the Cain–Abel struggle.

Status and distribution

Uncommon and local summer breeding residents in deciduous woodlands from the Baltic States, Belarus, western Russia, Poland, and Germany south to the Ukraine, and spottily farther south and east to Greece, Turkey, and the Caucasus. Their range was formerly much larger, but the loss of breeding habitat, shooting and other persecution, and egg collecting have eliminated them as breeding birds.

The total population is completely migratory; their winter range is entirely within sub-Saharan Africa.

Lesser Spotted Eagles are encountered in great numbers on migration between the summer and winter ranges primarily around the eastern end of the Mediterranean Sea. Over 150 000 have been counted in Israel during autumn migration.

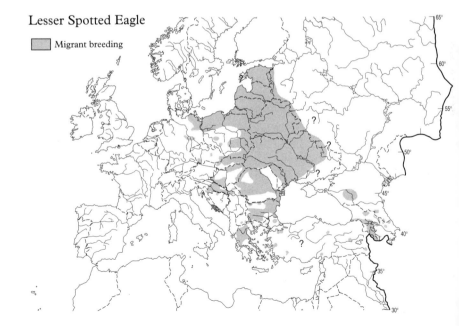

Lesser Spotted Eagle

☐ Migrant breeding

They have been recorded as vagrants in Spain, France, Sweden, Denmark, and the Low Countries, and are recorded in the tens regularly in spring at Cape Bon, Tunisia.

Fine points

Separating the spotted eagles in the field is often difficult, and some individuals cannot be identified for certain. Greaters often have white areas on their tarsi feathers, a feature not found so far on Lessers. Lesser juveniles show a rufous nape patch, thought to be diagnostic, but several juvenile Greaters have had the same patch, so this field mark is only suggestive of Lesser.

Subspecies

Aquila p. pomerina occurs in the Western Palearctic. *A. (p.) hastata* of India differs considerably and is most likely a separate species.

Etymology

Pomarina was a former dukedom of Poland, presumably where the first specimen was taken.

Measurements

Length: 54–62 cm (58 cm)
Wingspread: 145–165 cm (155 cm)
Weight: 1.2–2.2 kg (1.6 kg)

Greater Spotted Eagle   *Aquila clanga*

(Spotted Eagle)

Plates 29, 30; Photos, p. 332

Description

A medium-sized dark eagle that is a summer breeding bird in the eastern (primarily north-eastern) Western Palearctic. In all plumages they appear similar to but usually darker than Lesser Spotted Eagles. Both spotted eagles have rounded nostrils (oval or elliptical on other eagles) and short, tightly feathered tarsi (longer and looser in other eagles). The underwings of Greater Spotted Eagles usually show secondary coverts darker or the same colour as the flight feathers, but occasionally also coverts paler than flight feathers. Lesser Spotted Eagles almost always show coverts paler than flight feathers. When perched, Greaters' heads appear noticeably larger than those of Lessers; their longer nape feathers give their heads a more ragged appearance. Many Greater Spotted Eagles show white areas on lower tarsi feathers. Compared to other *Aquila* eagles, the two spotted eagles are noticeably smaller and, in flight, show relatively shorter head projection, wings, and tails. Almost all Greaters show the white *Aquila* patch at base of primaries, and a white wrist comma is usually visible on underwings. Wingtips reach or almost reach tail tip on perched eagles. Tail is dark brown—on some eagles becoming paler towards the tip with several narrow dark bands—with a wide pale terminal band. Eyes are dark to medium brown, lightening a bit with age. Beak is two-toned, dark at tip and horn-coloured at base. Cere, gape, and toes are yellow to pale yellow. The rare pale-colour morph is usually referred to as 'fulvescens' and is described after the normal morph.

Adult. Overall dark brown to blackish-brown (but occasionally medium brown), except for paler brown secondary wing coverts of some adults, white tips on greater uppertail coverts, absent on some (older?) adults, and pale tips of tail feathers. Undersides of flight feathers are greyish and usually appear paler than or the same colour as the underwing coverts, but rarely darker.

Juvenile. Head is dark brown, usually with buffy tips on nape feathers, and, rarely, a rufous nape patch like those of juvenile Lesser Spotted Eagles. Dark brown backs and upperwing coverts usually have pale short streaks or spots on tips; two rows of larger spots are noticeable on tips of median and greater coverts. The amount of spotting is variable from little to much (the name 'spotted' is derived from these). Greater uppertail coverts are white and form a pale 'U' at base of upper tail; other uppertail coverts and rump are dark brown, often with wide buffy streaking. Underparts and upper leg feathers are dark brown (occasionally medium brown) with buffy streaking. Streaking is heavier on breasts of some juveniles, sometimes forming a large pale breast patch; streaking is heavier on bellies of others; and yet others have uniformly buffy bellies. Undertail coverts are whitish, often with dark spotting. Tarsi feathers are dark brown, often with white areas near toes. Underwings show brown to dark brown lesser and median coverts, paler greyish-brown median coverts and flight feathers, the latter with faint narrow dark brown barring, usually with a wide unbanded area on tips. Inner primaries are somewhat paler on many juveniles.

Second winter eagles appear almost identical to juveniles in that the new feathers are identical to the replaced ones, except for being darker and fresher. New secondaries and tail feathers have wider white tips and differ from the old ones whose tips are usually worn, resulting in uneven ragged trailing edge of wings. Pale spots on upperwing coverts are also often worn and now appear as short narrow whitish streaks.

Third winter eagles are transitional, with a mixture of new, adult-like and old immature feathers. New nape feathers lack pale tips. New upperwing coverts are dark brown lacking white spots; old ones show at most small white spots or short streaks. New secondaries are darker and usually unbanded; if banded, then with faint, narrow pale bands. They are also longer than old secondaries; trailing edge of wing appears ragged as a result. New greater uppertail coverts are white; white 'U' at base of tail is still noticeable. New undertail coverts are white with dark barring or mottling.

Fourth winter eagles are essentially adults, but some may retain a few immature feathers. New uppertail coverts are dark brown with whitish tips. New undertail coverts are all dark brown.

Fulvescens-**morph juveniles** are rufous to rufous-buff on head, upper back, underparts, underwing coverts, and leg feathers in fresh plumage. But colour fades, often to creamy, by the next spring. Flight and tail feathers are as in normal morph, except that they usually lack any narrow dark barring. Older immatures are darker rufous, with many blackish streaks on face and body. Lower back, scapulars, and carpal patches are blackish. No specimens of adults of this morph were found, nor were there reports of adults breeding in the morph; it is possible that the colour of this morph is age-related so that the adult plumage of *fulvescens* is not separable from that of normal adults.

Unusual plumages

Hybrids with Lesser Spotted Eagles have been reported. If true, then the offspring would show characters of both. No other unusual plumages have been reported.

Similar species

Lesser Spotted Eagles (Plates 28, 30) are similar to Greater Spotted Eagles in all plumages but are usually paler and smaller (but with some size overlap), with narrower wings and longer tails. Their underwing coverts are usually paler than the flight feathers; opposite for most Greater, but both can show this, as well as uniformly coloured underwings. Head appears a bit smaller on perched Lessers, with a smoother, less ragged appearing nape. Lessers lack white areas on tarsi feathers. Adult Lessers have yellow eyes, compared to brownish ones for Greaters. Adult Lesser's brown undertail coverts have broad pale tips, not found on adult Greaters. Lesser juveniles usually have fewer and smaller spots on upperwing coverts.
Steppe Eagle (Plates 31, 32) adults appear similar in flight to adult Greater Spotted Eagles but show a wide dark tailing edge and dark carpal patches on underwings and have relatively longer wings and tails. In addition, they always have a rufous nape patch.

Eastern Imperial Eagle (Plate 34) adults appear similar in flight to adult Greater Spotted Eagles. See under that species for distinctions.

White-tailed Eagle (Plate 11) immatures can appear similar in flight to Greater Spotted Eagle. See under that species for distinctions.

Flight

Powered flight is with rather deep deliberate wingbeats of somewhat cupped wings. Soars with wings held level or with wingtips a bit depressed. Glides with wings level to wrist or with wrists above body and with wingtips depressed deeply.

Moult

They apparently undergo an incomplete annual moult, in that not all flight feathers (perhaps not all body feathers, as well) are replaced every year, but detailed studies are lacking. Moult is most likely commenced in early spring and suspended for the winter in late autumn; however, breeding adults may suspend their moult for a period.

Behaviour

Opportunistic predators, taking a wide range of prey, including small mammals, birds (mainly water birds but also nestlings), reptiles, amphibians, insects, and fish. They regularly pirate prey from other raptors, and, in winter they also feed on carrion.

Hunting is mainly from perches, but also on the wing, and, occasionally, by walking on the ground. They are usually somewhat sluggish and take easily captured prey, but they can be aggressive and dashing, occasionally taking prey as agile as pigeons.

Flight displays include undulating flight and mutual soaring. Typical raptor stick nests are built on trees, and one to three eaglets are raised.

Status and distribution

Uncommon and local summer breeders in riparian and wet woodlands, never far from water and usually at lower elevations, through-

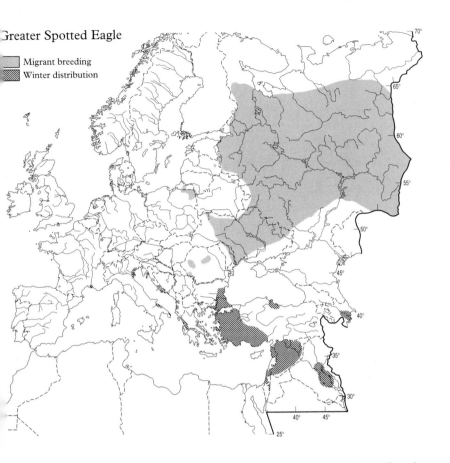

Greater Spotted Eagle

Migrant breeding
Winter distribution

out the north-eastern Western Palearctic, mainly in northern Russia but also west to southern Finland (barely), Sweden (one breeding record), Estonia, Latvia, Lithuania, Poland, Belarus, the Ukraine, and Romania.

They are completely migratory, and their winter range in the south-eastern Western Palearctic includes Italy, Turkey, Jordan, Iraq, Israel, Egypt, and northern Saudi Arabia. Although usually found near water, they also occur in arid areas away from water on occasion. They are encountered on migration between the summer and winter ranges.

Vagrants are regularly recorded on migration in Scandinavia and Holland.

Fine points

Separating the spotted eagles in the field is often difficult, and certain individuals cannot be identified for certain. Greaters often have white areas on their tarsi feathers, a feature not found so far on Lessers. Lesser juveniles show a rufous nape patch, thought to be diagnostic, but several juvenile Greaters have had the same patch, so this field mark is only suggestive of Lesser.

Subspecies

Monotypic.

Etymology

'Spotted' for the markings on the upperwing coverts of juvenile and first summer eagles; *clanga* is from the Greek *klangos* for a kind of eagle (mentioned by Aristotle).

Measurements

Length: 57–65 cm (61 cm)
Wingspread: 159–175 cm (167 cm)
Weight: 1.4–3.2 kg (2.1 kg)

Tawny Eagle   *Aquila rapax*

Plates 32–34, 47

Description

A rare and local resident in Morocco and a vagrant in Israel. Most birds have some tawny coloration; however, it is a polymorphic species, with much individual variation from creamy-white to uniformly dary brown but with little age variation. The palest and dark brown birds are the only ones to lack tawny coloration in fresh plumage, but tawny coloration fades in time to creamy or buff.

Tawnys have a classic aquiline build, with well proportioned body and wings. Like most *Aquilas*, flying birds usually show pale areas at base of primaries and pale area at base of tail on uppersides; inner primaries are noticeably paler than other flight feathers. All but darkest birds show two-toned upperwing in flight; pale coverts contrast with darker flight feathers. Baggy 'trousers' are noticeable on perched birds; wingtips reach tail tip. Sexes are alike in plumage; females are larger and perhaps on average darker. Cere, gape, and legs vary from yellow to orange-yellow. Dark beak is horn-coloured at base. Tails are dark brown, usually with faint greyish banding on older and pale tips on juvenile and first-year birds.

Age variation in Tawny Eagles is mainly in eye colour and pattern on secondaries; however, adults may be on average somewhat darker than younger birds. Older immatures have a hooded plumage not found in juveniles or adults. Plumages vary from pale creamy to dark brown, but only the dark birds need to be described separately. Moult of most is prolonged and may be occurring at any time; birds in transition can appear quite unusual, with dark heads and pale bodies or vice versa.

Light morph

Light morph eagles have uniformly coloured heads, backs, upperwing and uppertail coverts, and underparts; these vary from pale to dark tawny, with breasts either clear or with variable amounts of dark brown or dark rufous streaking. However, some are creamy, whitish, or greyish-brown, lacking any rufous coloration. Those with dark breast streaking usually have darker backs as well. Underwing coverts are usually same colour as belly and contrast with dark flight feathers.

Light-morph adults have yellow eyes and grey secondaries with dark banding, with dark band on tip not noticeably wider than others. Scapulars are dark brown. Compared to younger birds, they average more dark brown on lower back and upperwing coverts.

Light-morph juveniles have dark brown eyes and uniformly unbarred grey-brown secondaries with wide white band on pointed tips. Tail has narrow pale terminal band.

Light-morph second year birds are like juveniles, except that secondaries are less pointed and blunter but still with pale tips. Secondaries are sometimes faintly barred, but barring is not noticeable in field. Some birds in this age class also appear 'hooded'.

Light-morph third year birds often show darker heads and upper breasts, forming a hood. Other birds are somewhat dark on breast and back, but with paler tawny head. Eye is yellow-brown to pale brown and secondaries are grey with some faint narrow dark barring.

Light-morph fourth year birds are like adult, but eyes vary from yellow to darker yellow-brown and barring on secondaries is narrower, especially outermost bar.

Dark morph

Dark morph eagles' secondaries and eye colour vary with age as in light morphs.

Dark-morph adults are usually uniformly dark brown to blackish-brown.

Dark-morph non-adults from juvenile to fourth year are also dark, but usually with paler belly, legs, and undertail coverts. Many show pale spots on feathers of upper back, nape, and underparts.

Note: Tawny Eagles were thought to be conspecific with Steppe Eagles because of their similar plumages, but they are separable in the field and are, in reality, quite different.

Unusual plumages

A photograph of a partial albino adult was taken in Kenya.

Similar species

Steppe Eagles (Plates 31, 32) can appear similar in plumage to many Tawny Eagles. Steppes are generally more sluggish and lethargic than are Tawnys, and appear more horizontal and elongated when perched. Adult Steppe Eagles are overall dark brown like darkmorph Tawnys, but can be separated by rufous nape patch and darker eyes, and have blackish carpal patches and wide dark terminal band on undersides of secondaries. Steppes share two other plumages with Tawnys—those with underparts grey-brown or tawny (rare in Steppe). In both of these, Steppe is distinguished by white band on underwing formed by white underwing coverts and by sharp line of contrast between white undertail coverts and dark belly; neither are found on Tawnys. The white tips on secondaries of Steppes are also wider than those on Tawnys. See also 'Fine points' in Steppe Eagle account.

Eastern Imperial Eagle (Plate 34) juveniles and second winter eagles can appear similar to same-age Tawny Eagles but have darkly streaked breast and are larger.

Spanish Imperial Eagle (Plate 33) juveniles and second winter eagles appear almost identical to same-age Tawny Eagles but are somewhat larger. Distinction is not always possible.

Greater Spotted Eagles (Plates 29, 30) can appear similar to some Tawny Eagles. See under that species for distinctions.

Lesser Spotted Eagles (Plates 28, 30) can appear similar to some Tawny Eagles. See under that species for distinctions.

Flight

They are agile aggressive predators. Active flight is with strong, steady wingbeats of slightly cupped wings. Soar with wings slightly cupped; glide with wrists raised and primaries drooping.

Moult

Juveniles begin moult of tail and flight feathers when about a year old. Annual moults should be complete, but apparently do not include all secondaries. Birds should be moulting throughout the year, except during times of food stress or when breeding. Moult is not well studied. Adult plumage reached in four years.

Behaviour

Active predators that take a wide variety of prey from insects to large animals and birds. They are also accomplished pirates, stealing prey from many other predators. And they are not above eating carrion. Often seen chasing and, less often, capturing birds. Usually Tawnys are not social, but sometimes groups of non-adults gather at termite mounds or other sources of abundant food.

Status and distribution

Occur regularly in the Western Palearctic only in Morocco, with a few dozen pairs resident breeders in the flood plain of the Sous River. Formerly they were more widespread and bred in Algeria and Tunisia. A photograph of a vagrant was taken in Israel, where two more records are being considered. Two mislabelled specimens from Sardinia were recently identified as Tawnys, and there are sight records from Gibraltar.

Fine points

Can be separated from Steppe Eagles by gape size and plumages. See 'Fine points' under that species.

Subspecies

A. rapax belisarius is resident in North Africa.

Etymology

rapax is Latin for 'rapacious'.

Measurements

Length: 57–67 cm (62)
Wingspread: 157–190 cm (175)
Weight: 1.6–2.6 kg (2.1)

Steppe Eagle   *Aquila nipalensis*

Plates 31, 32; Photos, p. 334

Description

A large clumsy dark eagle that is an uncommon breeding bird of steppes and plains of extreme eastern Europe but migrates in numbers through the Middle East. Steppe Eagles are a bit smaller than Golden and Imperial Eagles, and in flight they also appear somewhat less agile than those eagles, even appearing floppy. Like many *Aquila* in flight, they usually show pale areas at base of primaries on uppersides and, less frequently, a white patch on lower back. They perch more horizontally than other *Aquila*, with lower body and tail giving them a somewhat elongated appearance. Wingtips reach the tail tip on perched birds. Sexes are alike in plumage; females are larger. Cere, gape, and legs vary from yellow to orange-yellow. Gape is noticeably wide; yellow mouth skin extends beyond centre of eye. Beak is black with pale area at base, larger on older birds. Tails are either uniform dark brown, often with fine greyish banding, or greyish with wide dark brown banding. Tails of juveniles and first summer birds have wide pale tips; tails of older eagles have dark tips. Adult plumage is acquired after four moults.

Adults are overall dark brown, including tail coverts, except for variably sized rufous nape patches, paler throats, and pattern on underwings. Underwings show black carpal patches, paler dark brown medium coverts, and greyish barring on undersides of dark flight feathers that stops short of the tips, forming a wide dark band on trailing edge. Underside of dark tail shows faint pale narrow bands and wide dark tip. Iris is yellow with brown flecking.

Juvenile plumage is distinctive and consists of uniform colour on head, back, coverts, underparts, and leg feathers, varying from light to medium to (rarely) dark brown, to greyish-brown, or (even rarer) tawny, except that on most individuals all (upper and under) greater wing coverts have wide white or creamy tips in fresh plumage. This shows as a *distinctive white band across centre of underwings and upperwings* of flying birds. Greyish secondaries and inner primaries likewise have wide white tips that show as an even-width pale band on trailing edge of wing. Uppersides of flight feathers are dark brown. Occasional birds have greyish-brown tips to greater wing coverts and as a result do not show the pale bands through centre of wings; others show only pale bands as a white 'L' around the carpal patches of underwings, but all have wide white secondary tips. Greater uppertail and all undertail coverts are white to creamy. Dark tail has wide pale band on tip. Iris is dark brown.

Second winter birds are essentially like juveniles, as replacement feathers are identical to juvenile feathers. Three to six replaced secondaries at three moult centres (S1, S5, and S14) appear longer than retained ones, which usually have white tips worn off (see 'Moult'). New secondaries have wide white tips and may show dark banding, resulting in ragged trailing edges of wings. White tips on greater upperwing coverts usually wear and are visible as a poorly defined narrow pale line, whereas underwing covert tips do not wear noticeably. Some replacement body and covert feathers may appear somewhat darker than the retained juvenile ones, giving a somewhat mottled appearance on some eagles.

Third winter birds differ from those in the previous two plumages in that new secondaries and tail feathers lack wide white or wide dark tips. The trailing edges of the wings now lack wide white tips on any feathers. New greater underwing coverts may have some dark areas but a well-defined band through the centre of the underwings is still noticeable. New body feathers are somewhat darker than those of previous plumages so that eagles of this age appear more mottled. Tail coverts are still overall pale.

Fourth winter birds are characterized by new secondaries that have wide dark subterminal bands and old ones with narrow dark subterminal bands, greater underwing coverts that are a somewhat even mixture of white and dark, and noticeably darker underparts. Pale tail coverts now show dark barring. Perched eagles of this age can appear much like adults, but their upperwing coverts appear paler than their backs and they may or may not show the pale rufous nape patch. The paler upperwing coverts can also be noticed on flying eagles, which often appear similar to adult Lesser Spotted Eagles (*A. pomerina*). Iris is medium brown.

Fifth winter birds are similar to adults, but tail coverts have some white markings and underwings show both new darker secondaries with wide dark terminal bands and old ones with narrow terminal band.

Unusual plumages

An adult specimen had many white feathers on upper back and belly.

Similar species

(*Note.* Steppe Eagles in first three plumages with white bands through wings are distinct from all other eagles.)

Lesser Spotted Eagles (Plates 28, 30) appear similar. For differences see under 'Similar species' of their account.

Greater Spotted Eagles (Plates 29, 30) appear similar. For differences see under 'Similar species' of their account.

Tawny Eagles (Plates 32, 47) of Africa appear similar. For differences see under 'Similar species' of their account.

Golden Eagles (Plate 35) fly with wings straight, often in a dihedral. All are mostly dark but with rufous to golden crown and nape. Adults and subadults have tawny bars across median upperwing coverts. Juveniles and subadults have white areas on base of tail and often at base of inner primaries and outer secondaries.

Eastern Imperial Eagles (Plate 34) show more protruding head and neck in flight and fly with wings held level. Adults are dark like adult Steppes but have much larger, straw-coloured nape patch, a few white scapulars, solid white undertail coverts, and are overall darker. Subadults can show a two-toned pattern or extensive white mottling on underparts. Juveniles are similar to young Steppes (particularly those without white bands through wings) but have streaked underparts, white area at upper base of tail extending on to lower back, and more contrast between darker secondaries and paler inner primaries.

Flight

Powered flight is with slow, laboured, floppy wingbeats. Soars with wingtips slightly curved downward. Glides with forewing level and wingtips noticeably drooped.

Moult

All birds undergo an incomplete annual moult in that they do not replace all body, flight, and tail feathers every year. They moult on both summering and wintering grounds but suspend moult during migration. Breeding adults may also suspend moult. Second winter eagles typically show ragged trailing wing edges caused by old, shorter and new, longer secondaries. Juveniles begin moulting at around 1 year of age. Adult plumage is acquired in 4 years.

Behaviour

Generally sluggish raptors, lacking the dash and *élan* of Golden or Eastern Imperial Eagles. They are dependent on ground squirrels (susliks) for food in summer and are often seen perching on ground waiting for them to emerge from burrows. They regularly pirate prey from other raptors.

Nest is usually placed on the ground or in a small bush, even (rarely) on haystacks. They often eat carrion, particularly during spring migration.

Status and distribution

Breeds in small numbers in the steppes of eastern Europe between 42° and 52°N and east of 41°E. Entire population leaves breeding area for winter, most going into the Arabian peninsula or Africa south of Sahara. A few winter in the Middle East. Many more birds from central and western Asia migrate through this area, even more during spring. On migration encountered singly or in small groups away from concentration areas, using thermals to migrate. Large concentrations have been recorded at Elat, Israel and Suez, Egypt in spring. They occur widely as a vagrant in northern Europe and Italy.

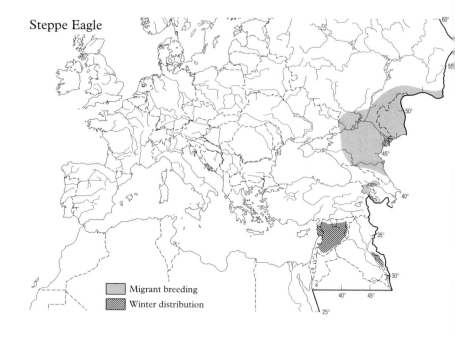

Steppe Eagle

Migrant breeding
Winter distribution

Fine points

Can be distinguished from Tawny Eagles by gape size; the mouth skin extending back past the centre of the eye on Steppe, but to the centre on Tawny. But this should be used with caution, as the relative positions of eye and gape can appear to be different depending on the viewing angle.

Subspecies

Aquila nipalensis orientalis occurs in Western Palearctic. Eastern race *nipalensis* differs only in clinally larger size.

Etymology

nipalensis for Nepal, where apparently the first specimen was collected.

Measurements (for race *orientalis*)

Length: 60–77 cm (69 cm)
Wingspread: 163–205 cm (186 cm)
Weight: 1.8–3.8 kg (2.7 kg)

Spanish Imperial Eagle   *(Aquila adalberti)*

Plate 33; Photos, p. 336

Description

A large, robust dark eagle, an uncommon resident in and endemic to central and south central Spain. They are similar in plumage and behaviour to the Imperial Eagle, *A. heliaca*, and were formerly considered a race of that species. Like that species, they show longer head and neck projection and their wing shapes resemble those of *Haliaaetus* eagles. Wingtips fall just short of tail tip on perched eagles. Adult plumage is attained after four annual moults. Eye is straw-coloured, with a fine, close-grained brown flecking over inner and lower eye. Beak is two-toned: dark on tip and horn-coloured on base. Cere and gape are yellow. Feet are yellow.

Adult. Overall blackish-brown, except for large straw-coloured patch on crown (but not forehead), nape, hind-neck, and cheeks; extensive white spotting on upper scapulars, marginals, and lesser upperwing coverts; and mottled undertail coverts. *Dark throat contrasts with straw-coloured cheeks. Leading edges of wings (marginal coverts) are white*, noticeable on flying adults. Undersides of wings are uniformly dark, except for narrow white bar on leading edge. Grey tail has narrow dark banding and wide dark terminal band.

Juvenile. Head, body, and wing and tail coverts are rufous, except for a few blackish scapulars on lower back. Greater upperwing coverts are dark brown with rufous tips that form a narrow pale band across upperwings. Flight feathers are dark brown, except for three grey inner primaries. Dark unbanded secondaries have wide white tips, forming a pale terminal band. Unbanded tail is grey-brown with a wide pale tip.

Second winter eagles are almost identical in plumage to juveniles except that new darker rufous feathers contrast with old faded ones and underparts show pale streaking. Frosted uppersides of new secondaries and tail feathers appear paler with less abraded tips, compared to worn retained juvenile ones; trailing edges of wing and tips of tail appear ragged.

Third winter eagles appear overall mottled dark and pale, a mixture of new dark adult feathers and retained pale ones. *Throat is usually completely dark, contrasting with pale cheeks.* New secondaries and tail feathers have wide dark tips. New dark body feathers are first replaced on breast rather than belly and lower rather than upper back.

Fourth winter eagles appear much like adults but have pale mottling throughout body and wing coverts. Straw-coloured crown and nape and white undertail coverts are easily noticeable.

Fifth winter eagles are first plumage adult. Some may show a few retained pale body or covert feathers.

Unusual plumages
No unusual plumages have been described.

Similar species
Golden Eagle (Plate 35) can appear similar to adult Spanish Imperial Eagle. See under that species for distinctions.
Tawny Eagle (Plate 33) juveniles and second plumage eagles can appear almost identical to same-age Spanish Imperial Eagles. See under that species for distinctions.

Flight
Less agile in flight compared to Golden Eagles. Powered flight is with deep, rather heavy wingbeats. Soars with wings in a slight dihedral; glides on cupped wings, with wrists up and wingtips down.

Moult
They undergo an incomplete annual moult, actively moulting from March into November. They only replace some of their flight, tail, covert, and body feathers each year. Breeding adults may also suspend moult for a time. Juveniles begin their first moult in late spring and moult in the typical accipitrid pattern (see 'Moult' in 'Introduction'). Adult plumage is attained after four moults.

Behaviour

Powerful predators that prey mainly on small- to medium-sized mammals, particularly hares and rabbits, but also take birds, occasionally reptiles, and carrion. They also pirate prey from other raptors. They usually hunt from perches, but occasionally on the wing, and take prey on the ground. They are not agile enough to take most birds in the air.

Flight displays include undulating flight and mutual soaring. Typical raptor stick nests are built on trees, and one to four eaglets are raised annually.

Status and distribution

Uncommon local residents in wooded savannahs and sparse forests of central and south central Spain.

They formerly ranged throughout much of the Iberian peninsula and north-west Morocco, but their populations have been greatly reduced and their range diminished due to direct human persecution, poisoning, and conversion of forested areas to cultivation. There is a recent breeding record for Morocco.

Non-adults disperse away from natal areas, travelling as far as Morocco and Senegal.

Fine points

Both Golden Eagles and adult Spanish Imperial Eagles have pale crown and nape patch. But Spanish Imperial Eagles are readily dis-

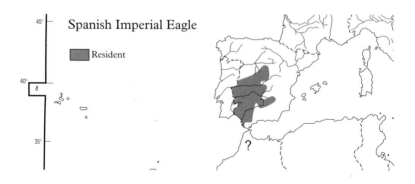

Spanish Imperial Eagle

Resident

tinguished by their larger head and beaks, longer necks, and paler colour of the patch, which extends further down on to the cheeks and contrasts more strongly with the Imperial's darker throat.

Subspecies

Monotypic. Most authorities now regard the Spanish Imperial Eagle, *A. adalberti*, as a separate species rather than as a race of *A. heliaca* as before.

Etymology

adalberti named in honor of Admiral Prince Heinrich Wilhelm Adalbert of Prussia.

Measurements

Length: 68–83 cm (75 cm)
Wingspread: 180–220 cm (200 cm)
Weight: 2.4–4.0 kg (3.2 kg)

Eastern Imperial Eagle   *(Aquila heliaca)*

(Imperial Eagle)

Plate 34; Photos, p. 338

Description

A large, robust dark eagle, a rare to uncommon local summer breeder of the steppes, open plains, and foothills of the east central Western Palearctic and rare local resident in south-eastern Europe and Turkey. Compared to other *Aquila* eagles in flight, they show a longer head and neck projection; this and the wing shapes of adults and juveniles resemble those of *Haliaaetus* eagles. Wingtips fall just short of tail tip on perched eagles. Adult plumage is attained after four annual moults. Eye is straw-coloured, with a fine, close-grained brown flecking over inner and lower eye. Beak is two-toned: dark on tip and horn-coloured on base. Cere and gape are yellow. Feet are yellow.

Adult. Overall blackish-brown, except for large straw-coloured patch on crown (but not forehead), nape, hind-neck, and cheeks; white spotting on middle scapulars; and pale undertail coverts. *Dark throat contrasts with straw-coloured cheeks.* Undersides of usually uniformly dark secondaries can show narrow greyish barring, with a wide dark terminal band, most likely on younger adults. Grey tail shows numerous narrow dark bands and a wide dark terminal band.

Juvenile. Head, underparts, and underwing coverts are creamy-buff and covered from the neck down with wide dark brown streaks, *which end abruptly in a straight line on the upper belly*—lower belly, leg feathers, throat, and undertail coverts are unstreaked. Back and lesser and median upperwing coverts are dark brown with wide creamy-buff streaking; greater upperwing coverts are dark brown with whitish tips that form a narrow pale band across upperwings. Flight feathers are dark brown, except for three grey inner primaries. Secondaries have wide white tips, forming a pale terminal band, but can show faint narrow pale banding. Uppertail coverts and lower back are whitish. Tail is brown with a greyish cast to upperside that fades on worn feathers, and has faint narrow dark bands.

Second winter eagles are identical in plumage to juveniles except for a more greyish cast to new feathers on uppertail. Uppersides of new longer secondaries and tail feathers appear paler with less abraded tips, compared to worn retained juveniles ones; trailing edges of wing and tips of tail appear ragged. Narrow pale band across upperwings (tips of greater coverts) is now faint or absent, but retained, faded median upperwing coverts form a wide pale bar across upperwings.

Third winter eagles appear overall mottled dark and pale, a mixture of new dark adult feathers and retained pale ones. *Throat is usually completely dark, contrasting with pale cheeks.* New secondaries and tail feathers have wide dark tips. New dark body feathers are first replaced on breast rather than belly and lower rather than upper back.

Fourth winter eagles appear much like adults but have some pale mottling throughout body and wing coverts. Straw-coloured head patch and white undertail coverts are easily noticeable.

Fifth winter eagles are first plumage adults. Some may show a few retained pale body of covert feathers. Narrow greyish barring on undersides of secondaries is usually rather bold in younger adults.

Unusual plumages

No unusual plumages have been described.

Similar species

Golden Eagle (Plate 35) can appear similar to adult Eastern Imperial Eagle. See under that species for distinctions.

White-tailed Eagle (Plate 11) immatures can appear similar to adult and older immature Imperial Eagles but have shorter, more wedge-shaped tails, white axillaries and band through underwings, and lack white undertail coverts and straw-coloured crown and nape.

Steppe Eagle (Plates 31, 32) can appear similar to adult Eastern Imperial Eagle. See under that species for distinctions.

Lesser Spotted Eagle (Plates 28, 30) is superficially similar to Eastern Imperial Eagle, but its smaller size, shorter head and tail projections, and relatively shorter wings readily distinguish it.

Greater Spotted Eagle (Plates 29, 30) is superficially similar to Eastern Imperial Eagle, but its smaller size, shorter head and tail projections, and relatively shorter wings readily distinguish it.

Flight

Less agile in flight than Golden Eagles but more agile than Steppe Eagles. Powered flight is with deep, rather heavy wingbeats. Soars with wings held level or in a slight dihedral; glides on cupped wings, with wrists up and wingtips down.

Moult

They undergo an incomplete annual moult, actively moulting from March into November. They do not replace all of their flight, tail, covert, and body feathers each year. Breeding adults may also suspend moult for a time. Juveniles begin their first moult in late spring and moult in the typical accipitrid pattern (see 'Moult' in 'Introduction'). Adult plumage is attained after four moults.

Behaviour

Powerful predators that prey mainly on small- to medium-sized mammals but also take birds, occasionally reptiles, and carrion. They also pirate prey from other raptors. They usually hunt from perches, but occasionally on the wing, and take prey on the ground. They are not agile enough to take most birds in the air.

Flight displays include undulating flight and mutual soaring. Typical raptor stick nests are built on trees, rarely on cliffs, and one to three eaglets are raised.

Status and distribution

Rare to uncommon local summer breeders on the steppes, plains, and foothills of southern Russia, the Ukraine, Slovakia, Serbia, and Romania and permanent local residents in Bulgaria, Macedonia, Greece, and Turkey. In winter they are found locally in small num-

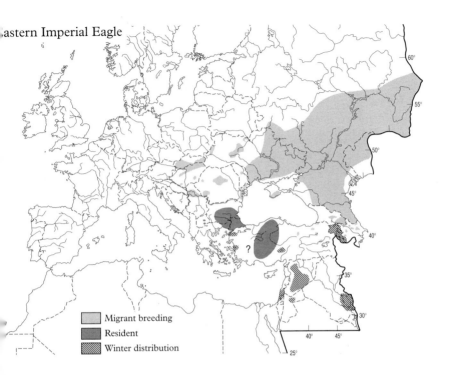

Eastern Imperial Eagle

Migrant breeding
Resident
Winter distribution

bers in Egypt, Israel, Jordan, Iraq, the United Arab Emirates, and Saudi Arabia.

Their populations have been greatly reduced in many areas due to direct human persecution and poisoning, but also due to conversion of open land to cultivation.

Fine points

Both Golden Eagles and adult Eastern Imperial Eagles have pale crown and nape patch. But Eastern Imperial Eagles are readily distinguished by their larger head and beaks, longer necks, and the paler colour of the patch, which extends further down on to the cheeks and contrasts more strongly with the Eastern Imperial's darker throat.

Subspecies

Monotypic. Most authorities now regard the Spanish Imperial Eagle, *A. adalberti*, as a separate species.

Etymology

heliaca comes from the Greek, *heliakos* meaning 'of the sun', referring to its nape colour.

Measurements

Length: 68–83 cm (75 cm)
Wingspread: 180–220 cm (200 cm)
Weight: 2.5–4.5 kg (3.2 kg)

Golden Eagle   *(Aquila chrysaetos)*

Plate 35; Photos, p. 340

Description

A large, powerful, agile dark eagle, uncommon to rare local breeding resident throughout most of the mountainous regions of the Western Palearctic. In all plumages they are distinguished by the *golden-tawny crown and nape*. In flight they appear longer-tailed and shorter-headed than do other dark eagles. *A tawny bar across each upperwing is noticeable* on all but juveniles. Wingtips fall somewhat short of long tail on perched eagles. Beak is two-toned: dark tip and horn-coloured base; cere and gape are yellow. Toes are yellow. Adult plumage is attained after four or five annual moults.

Adult. Head, body, and coverts are dark brown, except for tawny crown and nape, rufous upperleg feathers and undertail coverts, buffy tarsi feathers. Flight and tail feathers show greyish marbling, except on tips, resulting in a wide dark terminal band on underwings and undertail. Body and wing coverts appear somewhat mottled due to mix of old faded and new darker feathers. Median upperwing coverts are creamy to tawny; these form a pale bar visible on upper-wings of flying eagles. Some birds show whitish or rufous streaking on breast; a few adults have a few white scapulars. Eye is usually brownish-yellow sometimes with brown flecking, but occasionally brown.

Juvenile. Overall uniformly dark brown, darker than adult, almost blackish-brown in fresh plumage, except for tawny crown and nape and creamy-buff tarsi and undertail coverts. Paler juveniles may show some buffy streaking on breast and throat. Flight feathers are uniformly dark brown, lacking greyish marbling, with a variable amount of white on bases, resulting in white patches in the centre of underwings. However, some juveniles have no white on any of the flight feathers. The amount of white on the underwings varies among individuals and is not an age character. Tail has white base and dark tip; a few eagles have almost completely white tails. Eye is dark brown. Juveniles lack pale bars on upperwings and have more pointed secondaries.

Older immatures are usually distinguishable from juveniles and adults but are not easily classified as to age class in the field.

Second winter eagles are similar to juveniles but show mottled body and wing coverts due to mix of new dark and old faded feathers. Three to six inner primaries are usually replaced and appear fresher. New secondaries have wide dark tips; usually one or two inner and one or two outer ones are replaced, often number 5 as well, but also sometimes none (especially on northern eagles). New tail feathers are similar to those of juveniles but show greyish marbling in dark basal areas; usually central one or two and outer pairs are replaced in the first moult. The border between the white base and dark tips is less regular than on juveniles' tails. Pale band on upperwings are now noticeable.

Third and fourth winter eagles are intermediate between juveniles and adults in that they show white bases to some flight and tail feathers but otherwise appear much like adults. The white on the underwings now appears more as rays rather than patches. Eye is brown but can be brownish-yellow in some eagles.

Fifth winter eagles are essentially adult in plumage, but some may show a few retained immature flight or tail feathers with white bases. Some eagles at this age and older still have white on sides of some secondaries and tail feathers.

Unusual plumages

Both albinism and partial albinism have been reported.

Similar species

Eastern Imperial Eagle (Plate 34) adults are similar to Golden Eagles in being overall dark, with pale crown and nape and wide dark terminal tail band but are more blackish-brown, have noticeably longer head projection and white undertail coverts, fly with wings flat or cupped, and lack pale bar on upperwings and white patches on underwings. They show a well-defined line between pale cheek and dark throat; that of Golden Eagles is not well defined. White scapular patches are noticeable on back of flying and perched eagles.

Spanish Imperial Eagle (Plate 33) adults are similar to Golden Eagles in being overall dark, with pale crown and nape and wide dark terminal tail band but are more blackish-brown, have noticeably longer head projection and mottled undertail coverts, fly with wings in a slight dihedral or cupped, and lack pale bar on upperwings and white patches on underwings. When perched, large white scapular patches are noticeable.

Steppe Eagle (Plates 31, 32) is also a large dark eagle. See under that species for distinctions.

Flight

Agile, aggressive aerial predators. Powered flight is with fluid, strong deep wingbeats. Soars with wings held in a noticeable dihedral. Glides with wings held in a modified dihedral.

Moult

Golden Eagles undergo an incomplete annual moult, actively moulting from March into November. They do not replace all of their flight, tail, covert, and body feathers each year. Breeding adults may also suspend moult for a time. Juveniles begin their first moult in late spring and moult in the typical accipitrid pattern (see 'Moult' in 'Introduction'). Adult plumage is attained after four or five moults.

Behaviour

Swift, agile, and rapacious predators, taking a wide range of birds, mammals, and even lizards and snakes, also carrion, but preferred prey are rabbits and hares. They are capable of taking rather large prey, occasionally heavier than they are, and hunt on the wing, either in low-level rapid flight or from on high, but also from perches. Prey are usually taken on the ground, less often in the air. Golden Eagles have been reported dropping live turtles from a considerable height, no doubt to break the turtles open.

Golden Eagle

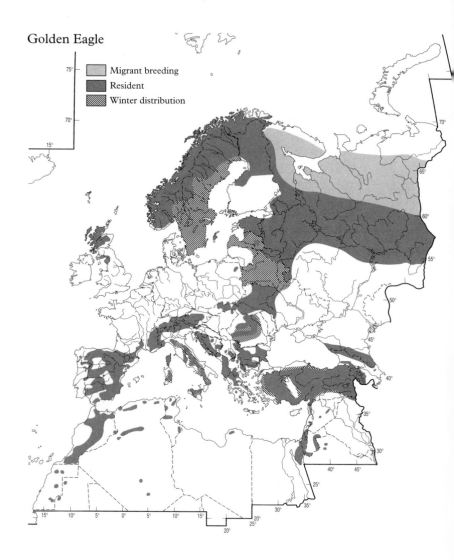

Migrant breeding
Resident
Winter distribution

Flight displays include undulating flight and mutual soaring. Typical raptor stick nests are built on cliff faces, where one to three eaglets are raised, but sometimes the old nests of a raven or other raptors on a tree or power pylon are used.

Status and distribution

Rare, sometimes uncommon, breeding residents of mountainous areas over much of the Western Palearctic, being absent only from the British Isles south of Scotland, western Europe from Sweden to southern France and Germany, and the extreme desert areas of northern Africa and the Middle East. Their populations have been somewhat reduced in many areas due to direct human persecution and poisoning.

Eagles from the northernmost part of the range migrate south for the winter. Immatures disperse away from breeding areas, particularly in winter.

Fine points

The colour of the nape does not change as Golden Eagles age. Napes of fledglings appear a somewhat darker orangish until faded by the sun some 1 or 2 months after fledging.

Subspecies

A. c. chrysaetos occurs throughout the Western Palearctic except for the Iberian Peninsula, north Africa, Turkey, the Caucasus, and the Middle East, where it is replaced by *A. c homeyeri*, which differs little. Variation is clinal, with northern eagles larger and paler, but with much individual variation in colour throughout.

Etymology

'Golden' refers to the colour of the crown and nape feathers; *chrysaetos* is from the Greek *chrysos* for 'golden' and *aetos*, 'eagle'.

Measurements

Length: 76–96 cm (86 cm)
Wingspread: 180–230 cm (205 cm)
Weight: 3.0–6.4 kg (4.5 kg)

Verreaux's Eagle   *Aquila verreauxii*

(African Black Eagle/Black Eagle)

Plate 48

Description

A large black eagle, a rare and local resident in open rocky, hilly and mountainous country of the Sinai peninsula and southern Israel and Jordan. The distinctive wing shape of pointed wingtips, bulging secondaries, and trailing edges pinched in at the body are useful for field identification. Plumages of the sexes are almost alike. Three plumages related to age are recognizable—adult, juvenile, and older immature. On perched eagles, wingtips reach or almost reach tail tip. Eyes are dark brown at all ages. Beak is horn-coloured at base and dark on tip. Cere, gape, lores, and eye-rings are bright yellow.

Adults are overall jet black, except for white 'V' on back, white rump and uppertail coverts, and primaries that are white with narrow dark barring and dark tips and form pale panels (windows) on black upper- and underwings. Toes are yellow to pale yellow.

Juveniles have a multicoloured plumage. Crown and upper nape are creamy-buff, throat and cheeks are black, and lower hind-neck and upper back are chestnut. Breast, back, and upperwing coverts are dark brown with wide buffy feather edges, resulting in a scaly appearance and almost white 'shoulder'. Belly, flanks, leg feathers, undertail and uppertail coverts, and rump are buffy with dark markings. Secondaries are greyish with marrow dark barring, and whitish primaries show narrow dark banding and dark tips on outer six and form pale primary panels (windows). Upperwing coverts are blackish with white feather edging. Tail is greyish with narrow dark banding. Toes are dull yellow.

Older immatures begin a slow continuous moult into adult plumage at around $1\frac{1}{4}$ to $1\frac{1}{2}$ years of age and finish when they are between 3 and 4. As a result, they appear intermediate in plumage between juveniles and adults, and are usually distinguished by the ragged trailing edges of wings.

Unusual plumages

Eagles with some white feathers have been reported.

Similar species

No other Western Palearctic eagle is similar.

Flight

Agile in flight, regularly hunting on the wing. They soar with wings held in a dihedral and glide with wings in a lesser dihedral or level.

Moult

Adults moult at a slow rate throughout the year, replacing all of their feathers in 2 years or so. Juveniles begin a very gradual moult into adult plumage beginning when they are about $1\frac{1}{4}$ to $1\frac{1}{2}$ years of age. Each feather replaced is adult, and this moult takes 2 years plus to complete. Unlike other large eagles, they have no discrete intermediate immature plumages.

Behaviour

Swift, powerful predators that specialize in catching rock hyrax. They also take some lagamorphs, small carnivores, large birds, and a few reptiles. They occasionally eat carrion but have not been reported pirating prey from other raptors.

Flight displays include undulating flight and mutual soaring. Typical raptor stick nests are built on cliff faces (rarely on trees on cliffs), where one eaglet is raised, as the older chick kills the younger in the Cain–Abel struggle.

They are usually silent except during the breeding season or when alarmed by predators.

Status and distribution

Rare and local breeding residents in several areas of the Sinai and one wadi in southern Jordan, perhaps elsewhere in the Middle East. Adults are seen regularly near Elat, Israel. There is an old nest record in northern Israel.

Verreaux's Eagle

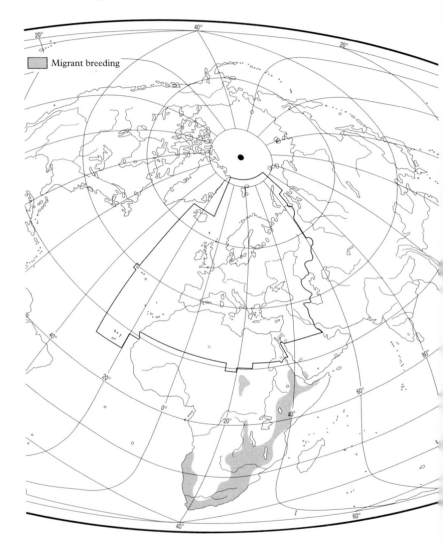

Migrant breeding

Fine points

Adults can often be sexed when seen from above while gliding, as males tend to hold their wings more extended so that their scapulars cover part of the white rump and separate the white 'V' in the back form the white uppertail coverts. Females usually show a continuous white patch, as do males in the hand or when landing. However, males usually have less extensive white feathering. Females have broader secondaries and usually fly with their tails more spread. Tail tips of females are more wedge-shaped than those of males.

Subspecies

Monotypic.

Etymology

Named after Jules Pierre Verreaux (1808–73), French naturalist, collector, and natural history dealer. Called Black Eagle in southern Africa.

Measurements

Length: 76–90 cm (84 cm)
Wingspread: 184–215 cm (199 cm)
Weight: 3.0–5.8 kg
       males: 3.8 kg
       females: 4.8 kg

Bonelli's Eagle   *Hieraaetus fasciatus*

Plate 36; Photos, p. 342

Description

A medium-sized pale eagle, an uncommon to rare local resident of the countries around the Mediterranean. The distinctive wing shapes of adults and juveniles are useful for field identification. Plumages of the sexes are alike. Four recognizably different plumages are age-related. On perched eagles, wingtips fall somewhat short of tail tip.

Adult. Head, back, and upperwing coverts are medium to dark brown, with a variable amount of white spotting on the back usually forming *a white triangular patch*. Uppertail is greyish to greyish-brown with narrow dark bands and a wide dark subterminal band. Underparts are white with narrow dark shaft streaks. Black median and greater underwing coverts form a black band across otherwise greyish underwings. Greyish flight feathers have indistinct whitish mottling, usually heavier on the primaries, but are unmarked on tips, which form a somewhat darker terminal band. Whitish leg feathers are heavily marked with brownish-grey barring or mottling and dark streaking. Undertail coverts are heavily mottled rufous-brown. Undertail is pale with wide dark subterminal band. Iris is yellow-orange; cere and toes are yellow to pale yellow.

Juvenile. Head is brown, somewhat paler than medium-brown back and upperwing coverts. Uppersides of primaries are a paler brown and contrast with darker brown secondaries. Uppertail is brown, with narrow, irregular dark bands. Breast is dark rufous in fresh plumage but fades to dull rufous-buff by winter and spring, and usually shows faint narrow dark shaft streaks. Belly, leg feathers, and undertail coverts are paler rufous and fade to creamy. Underwing coverts are rufous, and also fade to creamy; these often have black tips, which then form a narrow dark band through underwings. Primaries are whitish below, contrasting somewhat with light grey secondaries; both show narrow dark barring. Undertail is pale, with more or less distinct narrow dark bands. Iris is pale yellow-brown; cere and toes are yellow to pale yellow.

Second winter eagles are similar to juveniles in having brown upperparts and rufous underparts but their underparts and underwing coverts are a darker rufous and are marked with short bold streaks; their new flight and tail feathers have a greyish cast to uppersides and wide dark subterminal bands. Usually not all of the flight and tail feathers have been replaced; the number of each replaced is variable.

Third winter eagles are first plumage adults and differ from older adults in having wider dark streaks on their white underparts and paler, more noticeably barred undersides to flight feathers. The white area on the back may not be well developed.

Unusual plumages

No unusual plumages have been described.

Similar species

Booted Eagle (Plate 37) rufous-morph is similar in having rufous underparts and dark band through underwings. See under that species for distinctions.

Short-toed Snake Eagle (Plate 5) can appear similar. See under that species for distinctions.

Flight

Powered flight is rapid and agile with rather quick, powerful wingbeats of stiff wings. Soars with wings held flat, and glides with wings either flat or somewhat cupped.

Moult

Moult is not well known, but apparently they undergo a complete body moult but do not replace all flight and tail feathers every year. Post-juvenile moult begins in spring and moult of flight and tail feathers proceeds in the usual accipitrid pattern, with the inner seven primaries; inner four, outer two, and numbers 5 and 6 secondaries; and inner two pairs and outer pair of tail feathers usually replaced in the first moult.

Behaviour

Bold, swift, and agile predators, taking a wide range of birds, mammals, and even lizards and snakes. They hunt on the wing, either from on high or in low-level rapid flight, but also from perches. They are quite rapacious in pursuit of prey, which are taken on the ground or in the air and have even been reported to search for prey while walking on the ground.

Flight displays include undulating flight and mutual soaring. Typical raptor stick nests are built on cliff faces, where one to three eaglets are raised.

Status and distribution

Rare to uncommon, local permanent residents in hilly and mountainous regions, mainly in Spain and north-western Africa, but also sparsely in southern France, Sardinia, Sicily, Turkey, and the Middle East. Populations have declined because of direct persecution, electrocution, and poisoning.

Non-adult eagles disperse away from nesting areas, particularly in winter, and a few eagles may be seen at raptor migration concentration areas during migration.

Fine points

Some juveniles lack the narrow black band through the underwings.

Subspecies

H. f. fasciatus occurs throughout the Western Palearctic. The other race, *H. f. renschii*, occurs in the Lesser Sunda Islands and may actually be a separate species. The African Hawk-Eagle, *H. spilogaster*, had been considered a race of Bonelli's Eagle, but authorities now consider them species, based on their many differences, such as plumages, habitat preferences, and nest site.

Etymology

Common name after Professor F.A. Bonelli (1784–1830), an Italian naturalist. *Hieraaetus* comes from the Greek *hierax* for 'hawk' and

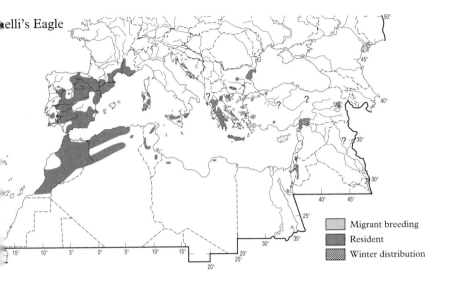

Bonelli's Eagle

Migrant breeding
Resident
Winter distribution

aetos for 'eagle'. *fasciatus* comes from the Latin *fascia* for 'band,' for the dark terminal tail band.

Measurements
Length: 55–75 cm (65 cm)
Wingspread: 145–180 cm (160 cm)
Weight: 1.4–2.4 kg (1.9 kg)

Booted Eagle   *Hieraaetus pennatus*

Plate 37; Photos, p. 344

Description

A small polymorphic, aerial eagle, a rare to fairly common summer breeding bird in the south-western, south-eastern, and east-central portions of the Western Palearctic. They occur in light, dark, and rufous colour morphs. Sexes are alike in plumage. Unique white spots or 'landing lights' are usually visible at the bases of the forewings, and all but the darkest eagles show pale inner primaries. All show pale bars across upperwings and a whitish 'U' at base of uppertail, as well. Wingtips fall several cm short of tail tips on perched eagles. Cere is bright yellow; bill is two-toned: horn-coloured on base and dark on tip. Toes are pale yellow to whitish. Adults and juveniles of all three morphs differ little and are not distinguishable in the field except for eye colour: golden-yellow in adults and dark to medium brown in juveniles, which changes gradually to adult colour by their first moult.

Light morph. Head is golden to pale brown with darker brown cheeks and whitish throat. Back is brown with pale areas on the scapulars, a feature not always visible. Upperwing coverts are brown, except for paler median coverts, which form a noticeable kite-like pale bar across each upperwing. Uppersides of flight feathers are a somewhat darker brown, except for pale inner three primaries. Underparts are whitish to creamy, with a variable amount of dark streaking, usually heavier on breast, and sometimes with a light rufous wash. Underwing and undertail coverts and leg feathers are whitish; each greater underwing covert usually has a dark spot near the tip. Undersides of flight feathers are darkish but show some darker barring when seen well under good light. Inner three primaries are decidedly paler. Tail appears dark from above, with white greater uppertail coverts forming a 'U' at their base. Underside of tail appears paler, usually with a more or less distinct dusky band on tip.

Dark morph. Head and upperparts are identical to those of light morph. Throat, underparts, and underwing coverts are dark brown, usually with narrow darker streaking visible on breast when seen in good light. Undersides of flight feathers are somewhat paler than coverts but still appear darkish. On darker eagles, the three inner primaries may not be noticeable paler than the others; similarly, their uppertail coverts may not be noticeably paler. Undertail coverts are dark rufous; undertail is same as that of light morph, but often appears reddish because of long undertail coverts. Leg feathers are buffy, contrastingly paler than dark underparts.

Rufous morph. Head and upperparts are identical to those of light- and dark-morph eagles. Throat, underparts, lesser underwing coverts, and leg feathers are light to medium rufous, with narrow dark streaking on breast. Paler rufous individuals fade to buffy in worn plumage. Median and greater underwing coverts are dark brown and form a wide dark bar across each underwing. Flight feathers are as in light morph, with inner three primaries paler. Undertail coverts are pale rufous and undertail as in other morphs. Some individuals appear intermediate between dark and rufous morphs.

Unusual plumages
Some dark-morph eagles show white spots on bellies and barring on leg feathers. One dark-morph individual was reported with pale buff uppertail. No cases of albinism or dilute plumages have been reported.

Similar species
Black Kite (Plates 8, 9) can appear similar to dark- and rufous-morph Booted Eagles. See under that species for distinctions.
Marsh Harrier (Plates 16–18) adult females and juveniles can appear similar to dark-morph Booted Eagles. See under that species for distinctions.
Egyptian Vulture (Plate 13) adults have the same colour pattern on their undersides as do light-morph Booted Eagles. See under that species for distinctions.
Bonelli's Eagle (Plate 36) juveniles and first summer birds also have rufous underparts and dark line through underwings and

appear similar to rufous-morph Booted Eagles, but they are larger, show pale flight feathers and narrower dark line through under-wings, and lack white 'landing lights' and dusky band on tail tip.

Flight

Powered flight is with deep, deliberate wingbeats of somewhat elastic wings. Soars with wings held slightly cupped, not flat, with tail somewhat spread. Glides with wings held in a kite-like cupped position, with wrists thrust forward and tail folded.

Booted Eagles are aerial hunters, soaring up to height and then gliding forward slowly while searching the ground directly below until prey is sighted, when they fold up to a tear-drop shape and hurtle downward, finishing the stoop with legs and talons outstretched. Sometimes they lower slowly on spread wings and tail from on high to view their prey better, then begin their stoop from a lower height. They do not hover.

Moult

They undergo a complete annual moult. They moult on both summer and wintering grounds but suspend moult during migration. Breeding adults may also suspend moult for a time. Post-juvenile moult begins late on spring migration and continues throughout the summer, but is most probably not completed until the next winter.

Behaviour

Aerial hunters, taking their prey of birds, mammals, and lizards on or near the ground or tree canopy, usually after a tremendous stoop from on high. If the prey is missed on initial stoop, they will often pursue it through the forest at high speed. They also hunt from perches and occasionally eat insects.

Their courtship displays have been described as the usual undulating flights but also high speed manoeuvring, usually accompanied by loud vocalizations. They are usually silent when not breeding.

They build a typical raptor stick nest in a large tree and raise one or two eaglets.

Status and distribution

Uncommon and local summer residents of the Western Palearctic, breeding in south-western Europe locally in France and throughout Spain and southern Portugal, north-western Africa from northern Morocco, Algeria, and Tunisia, and eastern Europe from Belarus, eastern Poland south through Ukraine, Slovakia, Hungary, Bulgaria, northern Greece, and south-eastern Asia in Turkey and the Caucasus.

Most of the population migrates south into sub-Saharan Africa for the northern winter, but a few winter in the Mediterranean Basin, with records from Israel, Egypt, southern Spain, Italy, and France, the Balearic Islands, Crete, Cyprus, and north-western Africa.

Has been recorded as a vagrant in England and almost all countries of continental Europe.

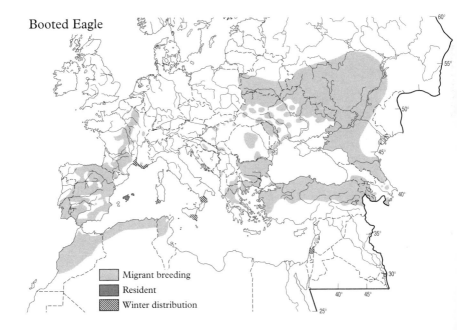

Booted Eagle

Migrant breeding
Resident
Winter distribution

Fine points

The undertails of dark-morph eagles appear reddish because of dark rufous undertail coverts.

Subspecies

Monotypic.

Etymology

Hieraaetus comes from the Greek *hierax* for 'hawk' and *aetos* for 'eagle'. *pennatus* is Latin for 'feathered' in reference to the completely feathered or 'booted' legs.

Measurements

Length: 42–50 cm (46 cm)
Wingspread: 113–134 cm (124 cm)
Weight: 555–965 g (735 g)

FALCONS—GENUS *FALCO*, FAMILY FALCONIDAE

Fourteen species of falcons occur in the Western Palearctic; 12 as breeding birds and two as vagrants. Five are characterized as 'large' falcons, of which there are two subgroups; first are the specialized bird-hunting Peregrine and Barbary Falcon, and second are the less specialized Gyrfalcon, Saker Falcon, and Lanner Falcon. All the rest are medium-sized, except for two small falcons: Merlin, a specialized bird hunter, and the vagrant American Kestrel. The 'Hobby' group is made up of three aerial insect specialists: the Eurasian Hobby and the local Eleonora's and Sooty Falcons. The latter two breed in late summer and feed their young on migrant birds. The 'Kestrel' group consists of two colonial insectivores, Lesser Kestrel and Red-footed Falcon, the generalist Common Kestrel, and the vagrant Amur Falcon.

Falcons belong in the family Falconidae, which is not close taxonomically to the other diurnal raptors in the family Accipitridae; nevertheless, the two share many characters, including sharp, curved talons, hooked beaks, keen eyesight, and predatory, piratical, and scavenging habits. Differences are mainly in structure, behaviour, and moult.

Falcons are characterized by long, pointed wings, dark eyes, and notched beaks. When excited, all bob their heads up and down and wag their tails up and down. Accipitrid raptors wag their tails from side to side.

Falcons do not build their own nests; instead they use cavities, cliff ledges, or potholes or appropriate stick nests of other raptors or corvids. Man-made structures, such as nest boxes, skyscrapers, or bridges are used by some species for nesting.

Falco and 'falcon' come from the Latin *falx* for 'sickle', for their wing shape in flight or, according to another source, for the shape of their beak or talons.

Lesser Kestrel   *Falco naumanni*

Plates 38, 39; Photos, p. 346

Description

A gregarious, slender, long-winged, medium-sized falcon, a widespread but local summer resident in southern and eastern Europe, north-west Africa, Turkey, and the western Middle East. Like the Common Kestrel, the two-toned pattern on upperparts is distinctive: reddish-brown or rufous back, outer secondaries, and upperwing coverts contrast with dark primaries. Main differences between these species are Lesser Kestrel's *pale talons, unmarked* (or lightly marked) *whitish underwing, lack of dark eye-lines, and wingtips reaching dark subterminal band of central tail feathers* on perched birds. Sexes have different plumages; females are only slightly larger. Juvenile plumages of both sexes are nearly alike and very similar to that of the adult female. Tail is wedge-shaped when folded. Central tail feathers are usually somewhat longer than the others; this shows as a projection on flying birds. While this is not diagnostic (a few Common Kestrels show it), it is suggestive, as almost all Lesser Kestrels show it. Iris is dark brown. Cere, eye-rings, and legs vary from orange on adult males to yellow on juveniles.

Adult male. *Head is bright blue-grey, with a creamy throat and lacking dark moustache marks and dark shaft streaking. Back and upperwing coverts are rufous and unmarked; greater upperwing coverts and inner secondaries are blue-grey.* Rufous-buff underparts are unmarked or lightly marked with fine dark spotting. Creamy undertail coverts and leg feathers are unmarked. Underwings show whitish unmarked flight feathers and whitish coverts that have little or no spotting. Uppertail coverts are unmarked blue-grey. Blue-grey tail is unbanded except for wide black subterminal band and narrow white terminal band, which is often worn off on central pair of feathers.

First summer males are males that have replaced juvenile body and some covert and tail feathers on the winter grounds. They are similar to adult males but appear more like adult male Common Kestrels, in that upperwing coverts that have not been replaced show heavy dark markings and head shows whitish cheeks and dark moustache marks *but lacks short dark eye-line and fine dark shaft streaks* on crown and nape. Underparts and underwing coverts are usually more heavily spotted than are those of adult males. Blue-grey tail usually has new longer adult male central tail feathers and older juvenile ones.

Adult female's head is reddish-brown, with fine dark shaft streaking, small whitish cheek patches, a thin dusky moustache mark under each eye, and white throat. Note *lack of short dark lines behind eyes*. Back and upperwing coverts are reddish-brown with dark brown triangular bars. Creamy buff underparts are finely streaked on breast, with spotting on belly and barring on flanks. Uppertail coverts are blue-grey, often with black barring, and contrast with reddish-brown tail, which may have a blue-greyish cast, but always has a wide dark subterminal band and many narrow dark brown bands.

Juveniles

Juveniles are similar to adult females and may not be distinguishable under field conditions; their backs are marked with even-width dark brown bands.

Juvenile male has greyish or cinnamon uppertail coverts like those of adult female and tail usually with a greyish cast.

Juvenile female has brownish uppertail coverts and brownish tail shows wider dark brown bands and lacks greyish cast.

Unusual plumages

A melanistic female was reported from Italy. Two adult females had adult male characters on one wing and one side of head; An adult female specimen had an adult male tail. A specimen of a second summer male had new tail feathers that were unusually patterned; they were grey with two black bands in lieu of the usual black subterminal band. One 2 cm wide band was approximately 5 cm from tail tip and the other was a 1 cm wide terminal band. Its outer tail feathers were retained juvenile ones.

Similar species

Common Kestrels (Plates 38, 39) are similar and occur throughout the range of this species. For differences see under Common Kestrel (see also 'Fine points' in this account.)

Red-footed Falcon (Plate 40) juveniles can also show a two-toned pattern on uppersides, but back and upperwing coverts are sandy brown, lacking any reddish coloration, and tails lack wider subterminal bands. Wingtips reach tail tip on perched birds.

Falcons of other species lack strong two-toned pattern of upperparts.

Flight

Powered flight is with light and buoyant wingbeats, deeper and slower than those of Common Kestrels. They soar on flat wings with tail somewhat fanned, glide on flat wings with wrists below body and tips curved up, and hover and kite frequently, as often as do Kestrels, but they are usually less persistent than that species in hovering in one place, moving to another hover location more frequently.

Moult

Complete annual moult is initiated after breeding, suspended during migration, and completed on winter grounds. Juveniles begin body moult on winter grounds; spring birds resemble adults. Some juvenile males apparently replace most or all tail feathers during winter.

Behaviour

Social and seldom encountered singly. Main prey are insects, but they also capture small snakes, lizards, frogs, mammals, and nestling birds. They are most active from late morning to late afternoon hunting over open or semi-open areas, and usually hunt in groups, both on the wing, from a hover, kite, or glide, or from conspicuous perches.

Nesting is colonial, with a preference for holes in buildings (including old castles) and cliffs, often in close association with humans. Occasional pairs, usually on range periphery, nest singly. Adults are vocal in nest colony, with 'keh, keh, keh' call most often heard, but they also utter Common Kestrel-like calls. Many birds leave the breeding colonies soon after young fledge.

Night roosts are formed throughout the year, with many birds roosting in one tree.

Status and distribution

Locally common near traditional breeding colonies throughout much of the southern part of the Western Palearctic, occurring primarily in open or semi-open areas. The range includes isolated areas of north-western Africa, most of the southern Iberian peninsula, southern France and Italy, Greece, Turkey, patchy areas in the Balkans, much of the western Middle East, and the Ukraine and southern Russia, where they occur much farther north than elsewhere.

Most birds move into southern Africa for winter; however, some sedentary individuals winter in southern Spain, southern Italy, including Sicily, and north-western Africa. A report of a falcon in January in central Italy is most unusual.

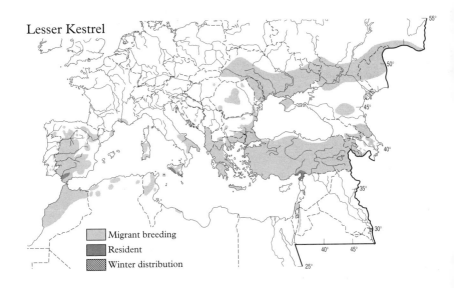

Lesser Kestrel

Migrant breeding
Resident
Winter distribution

Major population declines of Lesser Kestrels have been reported for many parts of the Western Palearctic, including extirpation in some countries. The decline is thought to be due to pesticides, loss of breeding sites due to building renovation and removal, intensification of agricultural practices, and loss of pastoral lands, both within the breeding range and in the African winter quarters. However, population increases in southern Spain have been noted.

They are not often seen at migration concentration locations. When seen, they are only observed in small numbers. However, large concentrations have been reported in parts of eastern Africa. Vagrants have been seen in Britain and central and western Europe, where they may be more common but overlooked because of their similarity in appearance to Common Kestrels.

Fine points

Lesser Kestrels have somewhat smaller heads than do Common Kestrels; as a result, they appear more hunched-over. Primary 9 is more than 6 mm longer than P8; less than 5 mm in Common Kestrel, but this is of limited field use.

Subspecies

None.

Etymology

Kestrel comes from the old French *crecerelle*, a name for *Falco tinnunculus*. Naumann was a nineteenth century German zoologist.

Measurements

Length: 26–31 cm (29 cm)
Wingspread: 66–72 cm (69 cm)
Weight: 100–160 g (130 g)

Common Kestrel *Falco tinnunculus*

Plates 38, 39; Photos, p. 348

Description

A common, widespread, long-tailed medium-sized falcon. In all plumages the *two-toned pattern on upperparts*—reddish-brown or rufous back, outer secondaries, and upperwing coverts contrast with dark brown primaries—distinguishes it in flight from all but Lesser Kestrel. Sexes have different plumages, but females are only slightly larger. Juvenile plumages of both sexes are almost alike and similar to that of adult female. Underwings appear pale in all plumages. Iris is dark brown. Cere, eye-rings, and legs vary from orange-yellow to yellow, paler on juveniles. *On perched birds wingtips fall short of dark subterminal band of central tail feathers*; noticeably long tail is wedge-shaped when folded.

Adult male. Head is blue-grey, with fine black shaft streaks and whitish cheeks, short dark lines behind the eyes, a thin dark moustache mark below each eye, and a white throat. Rufous back and upperwing coverts are lightly marked with small dark diamond-shaped spots. Creamy to buffy-rufous underparts are streaked on breast, spotted on belly and flanks. Creamy undertail coverts and leg feathers are unmarked. Uppertail coverts are uniform blue-grey. Blue-grey tail is usually unbanded except for wide black subterminal band but may have numerous narrow black bands on inner webs; most noticeable when tail is spread.

Adult female. Head is reddish-brown, with fine black shaft streaks, short dark lines behind each eye, whitish cheeks, a thin dark moustache mark under each eye, and a white throat. (Head can show a greyish cast on some (older?) females.) Back and upperwing coverts are reddish-brown with short dark brown triangular-shaped bars. Buffy underparts are streaked dark brown, tending to spotting on belly and barring on flanks. Creamy undertail coverts and leg feathers are usually unmarked but latter may show some fine short dark shaft streaks. Uppertail coverts vary from reddish-brown to blue-grey, usually with faint dark barring but may be uniform blue-grey without barring but always with narrow black shaft streaks. Reddish-brown to blue-grey tail has dark brown subterminal band and numerous narrow dark brown bands; many reddish-brown tails have a blue-greyish cast.

Juveniles are similar to adult females, but back and upperwing coverts have wide, even-width dark brown barring, and streaking on underparts is thicker and heavier. New adult feathers of the proper sex begin showing on back and uppertail coverts by first autumn; body moult is often completed by spring. Creamy leg feathers and undertail coverts are usually unstreaked, but former may show a few dark shaft streaks. Tail patterns of juvenile male are similar to those of the adult female; those of juvenile female are reddish-brown with noticeably wider dark brown bands.

Unusual plumages

A dilute-plumage adult male specimen exists. Albinism has been reported; most accounts are from Britain. Adult males with a rufous (not blue-grey) head and an adult female with a blue-grey (not reddish-brown) head have been captured or collected. An alleged melanistic Common Kestrel is illustrated in Harris *et al.* (1990).

Similar species

Lesser Kestrels (Plates 38, 39) of all plumages appear similar, but are smaller, more slender, and less heavily marked. Pale talons are best field mark, but are hard to see. Lesser Kestrels *lack short dark lines behind eyes*, and *wingtips usually extend into black subterminal band of central tail feathers* on perched birds. Tip of tail usually shows elongated central feathers; this is suggestive, however, but not diagnostic, because a few Common Kestrels also show this character. Gregarious behaviour and paler overall coloration are other suggestive aids. (See also 'Fine points' in Lesser Kestrel account.)

Lesser Kestrel adult males are distinctive and differ by having unmarked back and upperwing coverts, blue-grey greater secondary upperwing coverts and inner secondaries. Their heads are uniform blue-grey, except for creamy throats, lacking distinct moustache marks and whitish cheeks.

Lesser Kestrel first summer males are much more like adult male Common Kestrels, in that their retained upperwing coverts show extensive markings and they lack blue-grey colour on coverts and secondaries. They also show white cheeks, dark moustache marks, and heavily marked body and underwings. Best characters are pale talons, wing tip position of perched falcons, unmarked back, and lack of short dark eye-lines.

Lesser Kestrel adult females and juveniles appear very similar to same-age and -sex Common Kestrels and are separated only with care. Best characters are pale talons, wingtips reaching dark subterminal tail band on central feathers, lack of short dark eye-lines, and gregarious behaviour.

Red-footed Falcon (Plate 40) juveniles also show two-toned upperparts pattern, but back and upperwing coverts are sandy brown with no reddish coloration, and tails lack wide black terminal band. Wingtips reach tail tip on perched birds.

Falcons of other species lack the strong two-toned pattern on upperparts (but see under Saker Falcon).

Eurasian Sparrowhawks (Plate 23) can appear similar to Kestrels in flight. See under that species for distinctions.

Flight

Powered flight is with rapid, shallow, loose, almost fluttery, wing-beats, with wings travelling farther below body than above. Soars on flat wings with tail somewhat fanned. Glides on flat wings or with wrists below body and tips curved up. Hovers and kites frequently while hunting.

Moult

Annual moults are complete. Juveniles begin moulting body feathers in their first autumn, but flight feathers are not renewed until the next summer.

Behaviour

The often encountered Common Kestrel is usually seen hovering or sitting on an exposed perch, such as a pole, wire, or tree top, where they hunt for small mammals, birds, insects, lizards, or snakes.

They nest in cavities in cliffs or trees, crevices in buildings, or in old corvid nests. They readily use man-made nesting boxes, or even flower boxes. They are vociferous, and their easily learned calls carry some distance. In southern parts some breed communally, especially in colonies of Lesser Kestrels.

Status and distribution

Common and widespread throughout most of the Western Palearctic, occurring in all habitats except desert, tundra, and extensive forest; they prefer open habitats. North-eastern populations are migratory, with some birds moving well into sub-Saharan Africa.

Fine points

Some adult females (older birds?) appear like adult males, with blue-grey heads, tails, and uppertail coverts and rufous backs, but tail is usually heavily banded and grey uppertail coverts show black shaft streaks and markings on back are flattened triangles, not diamond-shaped as on adult males. Primary 9 is less than 5 mm longer than P 8; more than 6 mm on Lesser Kestrel, but this is of limited use under field conditions.

Subspecies

F. t. tinnunculus occurs throughout most of the Western Palearctic. *F. t. rupicolaeformis*, which hardly differs from the nominate, is resident in Egypt. Various insular races, which are somewhat different, reside on Atlantic islands.

Etymology

Kestrel comes from the old French *crecerelle*, a name for this species that derived from *crecelle*, meaning 'to rattle', for its call; *tinnunculus* is Latin for 'little bell-ringer', also for its call.

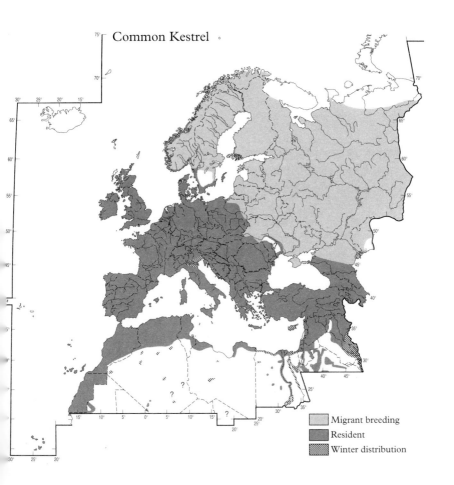

Common Kestrel

Migrant breeding
Resident
Winter distribution

Measurements

Length: 29–38 cm (34 cm)
Wingspread: 68–82 cm (76 cm)
Weight: 127–280 g (180 g)

American Kestrel   *Falco sparverius*

Plates 38, 39

Description

A small, colourful falcon that occurs as a rare vagrant in the Western Palearctic. Size and silhouette are same as those of male Merlin. Sexes have different plumages; juvenile plumage is almost like that of same-sex adult. Females average slightly larger and heavier, but there is considerable size overlap. Head consists of blue-grey crown, usually with a rufous crown patch, white cheeks and throat bordered by *two black moustache marks*, and *pale nape with black spots forming false 'eyes'*. Wingtips do not reach tail tip on perched birds. Iris is dark brown. Cere, eye-rings, and legs vary from orange to yellow, paler on juveniles.

Adult male. Back is rufous with black barring on lower half. *Upperwing coverts are blue-grey* with small black spots. Uppersides of primaries are black. Uppertail coverts are unbarred rufous, occasionally with some blue-grey and black feathers intermixed. Whitish to deep rufous breast is usually unmarked. Belly is paler with black spots, spotting heavier on flanks. White underwings show fine black markings. When wing is seen backlighted, *a row of white spots is noticeable on trailing edge of outer wing*. Unmarked leg feathers and undertail coverts vary from white to creamy. Typical tail is rufous with a wide black subterminal band and a narrow terminal band that is either white, rufous, blue-grey, or combination of these. Outer tail feathers are usually white with black banding. Tail pattern and colour vary considerably, and some tails are banded black, white, and grey, with little or no rufous.

Adult female. Head is like that of adult male but paler, with dark grey streaking in crown patch. Back and upperwing coverts are reddish-brown with dark brown barring; *upperwings appear uniformly coloured* on flying birds. Uppertail coverts are reddish-brown with narrow dark brown barring. Creamy underparts are heavily streaked reddish brown. Creamy to white leg feathers and undertail coverts are unmarked. Buffy underwings show brown markings. Tail is reddish-brown with numerous dark brown bands; subterminal band is noticeably wider.

Juveniles

Juveniles are similar to same-sex adults and undergo post-juvenile moult during late summer and early autumn (see 'Moult' in this account).

Juvenile male differs from adult male by having whitish breast with heavy black streaking, black barring on back extending up to nape, and rufous crown patch with black shaft streaks.

Juvenile female is virtually identical to adult female. However, some juveniles have tail with dark subterminal band same width or only slightly wider than other dark bands.

Unusual plumages

Complete albinos and birds with some white feathers have been reported. A specimen exists of a gynandromorph, showing plumage characters of both sexes, but sexed internally as a female.

Similar species

Common Kestrels (Plates 38, 39) are similar but are considerably larger, have only one dark moustache mark and proportionally longer tails, and show a contrasting two-toned pattern on upper-wings in flight. Adult males have blue-grey (not rufous) tails and rufous (not blue-grey) wing coverts.

Merlins (Plate 43), especially males, have a similar silhouette in flight, but have a heavier, more direct and purposeful flight. They also lack the rufous tails and double dark moustache marks of American Kestrels.

Other small falcons are separated by larger size, wingtips that extend to tail tip, and lack of double dark moustache marks.

Flight

Active flight is light and buoyant, but quite unlike that of the Common Kestrel; it is more like that of Lesser Kestrel, with deeper, slower wing-beats. (However, the food-begging flight of mated adult females and fledglings is similar to flight of Common Kestrel.) Soars on flat wings, often with tail spread. Glides on flat wings or falcon-like with wrists down and wingtips upswept. Hovers and kites regularly.

Moult

Annual moults are complete. Juveniles begin a more or less complete body moult in late summer to early autumn, but do not renew their flight and tail feathers until the next summer.

Behaviour

Behave much like Common Kestrels. They take, perhaps, more insects and fewer birds than does that species, but their staple is also small mammals.

Their calls are almost identical to those of the Common Kestrel.

Status and distribution

Have been reported from Scotland, England, Denmark, Estonia, Malta, and the Azores.

Fine points

American Kestrels possess a pair of 'false eyes', or 'ocelli', on the nape. This is thought to be protective coloration that serves to deter potential predators.

Subspecies

Falco s. sparverius occurs over most of North America and is migratory; this is the most likely origin of our vagrants as the other races are non-migratory.

Etymology

sparverius is Latin for 'pertaining to a sparrow', after this falcon's original common name of 'Sparrow Hawk'.

Measurements

Length: 22–27 cm (25 cm)
Wingspread: 52–61 cm (56 cm)
Weight: male 97–120 g (109 g)
     female 102–50 g (123 g)

Red-footed Falcon   *Falco vespertinus*   (Western Red-footed Falcon)

Plate 40; Photos, p. 350

Description

A gregarious, slender, long-winged small falcon. Adults have sexually dimorphic plumages; juveniles have yet a different plumage. Females are slightly larger. Iris is dark brown. Wingtips extend beyond tail tip on perched birds.

Adult male. Head, body, and wing coverts are dark grey, somewhat paler grey on undersides. Leg feathers and undertail coverts are bright rufous. Flight feathers are silvery above and contrast with dark grey coverts; below they are dark grey, contrasting somewhat with slightly paler coverts. Unbanded tail is blackish-grey. Cere, eye-rings, and legs are red-orange, with red-orange extending on to base of black beak.

Adult female. Crown and nape are dark orange-buff, cheeks are whitish, and throat is creamy. Small blackish areas around each eye extend down as short moustaches. Underparts are orange-buff, paler on the breast; some birds (younger?) have fine black shaft streaking on breast and flanks. Back, upperwing coverts, and flight feathers are blue-grey with black barring. Underwings appear pale, with orange-buff coverts and heavy whitish markings on flight feathers, with obvious wide dark band on trailing edges. Tail is blue-grey with black bands; subterminal band widest. Cere, eye-rings, and legs are orange, with orange extending on to base of black beak.

Juvenile. Forehead is whitish, and crown is brown; both are finely streaked with dark brown. Rest of head is whitish, with small black areas around and behind each eye that extend down to form small moustaches. Whitish nape forms a pale collar. Back and upperwing coverts are dark brown with greyish-brown barring and wide rufous feather edges. These appear sandy brown on distant falcons and contrast with dark brown flight feathers, forming kestrel-like two-toned upperwing pattern. Creamy underparts have wide dark brown streaking. Creamy leg feathers and undertail coverts are unmarked. Underwings appear pale, with some dark markings on creamy coverts and flight feathers and noticeable wide dark band on trailing edges. Tail consists of numerous even-width bands, varying from grey and black bands to rufous-buffy and dark brown bands with a greyish cast. Cere, eye-rings, and legs are yellow to yellow-orange, with colour extending on to base of beak.

Second summer falcons

Second summer falcons (juveniles returning in spring) of both sexes resemble respective adults, as they have undergone a more or less complete body moult, including some underwing coverts.

Second summer males usually appear much like adult males, except for retained flight and tail feathers, and usually have buff-orange on breast and nape and white throat and cheeks. Greyish underparts show fine black streaking. Central tail feathers, however, have usually been replaced. First adult tail usually shows a dark sub-terminal band. Cere, eye-rings, and legs are red-orange like those of adult male.

Second summer females appear almost identical to adult females and differ only in their fresher, darker body plumage with noticeable black shaft streaking on crown, nape, and underparts and retained juvenile flight and tail feathers. Retained juvenile underwing coverts are heavily marked. However, replaced central tail feathers have adult pattern and colour, noticeable in field.

Second winter falcons on autumn migration usually have some retained juvenile flight feathers, particularly noticeable on males as 'patchwork' primaries.

Unusual plumages

No unusual plumages have been reported.

Similar species

Eurasian Hobby (Plate 42) appears similar to juvenile Red-foot but is larger, has short pale superciliaries and less strongly banded tail, and uniformly coloured, not two-toned, upperwing, uniformly coloured underwings, and lacks pale collar. Adult Hobbys also have rufous legs and undertail coverts like adult male Red-footeds but have streaked undersides and strong face patterns.

Common Kestrel (Plates 38, 39) also has two-toned upperwing like juvenile Red-foot, but is larger, has heavier moustache and wide black subterminal tail band in all plumages. Perched Common Kestrels' wingtips fall short of tail tip.

Eleonora's Falcon (Plate 41) dark morph appears similar to adult male Red-foot but is larger with proportionally longer wings and tail and shows a pale line at base of flight feathers below. They also lack silvery uppersides of primaries, bright orange soft parts colour, and bright rufous leg and undertail feathers.

Sooty Falcon (Plate 42) adult also appears similar to adult male Red-foot, and juveniles of both are even more alike; see under 'Similar species' in that account.

Merlin (Plate 43) is somewhat similar. See under that species for distinctions.

Flight

Active flight is graceful, light and buoyant, with occasional quick wingbeats and longer periods of gliding, with constant adjustment of wings and tail. Soars on flat wings; glides with wrists below body and wingtips curved upwards. Hovers kestrel-like regularly but not persistently.

Moult

Adult moult is complete; it is begun in summer, suspended during migration, and completed in winter quarters. Juveniles replace most of body feathers and central tail feathers in winter before return migration. First year birds begin flight feather moult in summer and complete in winter quarters.

Behaviour

Social and only occasionally encountered singly, then usually as vagrants. They prey on a wide variety of insects but feed young mostly on small mammals, frogs, lizards, and nestling and fledgling birds. When actively pursuing aerial insects, flight is somewhat erratic with jerky movements on fast wingbeats. They also perch on ground and run to capture prey.

Red-foots usually nest in colonies, most often using unused nests in corvid breeding colonies, but also nest singly, again most often in old corvid nests. They form communal night roosts, except when breeding, and usually migrate in small to large flocks.

Status and distribution

Common in the south of their eastern European breeding range, less common in the north. However, local numbers fluctuate widely according to prey availability. Extraliminal breeding attempts have been recorded in France, Switzerland, Italy, Austria, Germany, Sweden, Finland, and Poland.

Entire population migrates to south-western Africa; autumn migration is mainly around the eastern Mediterranean and occurs in

Red-footed Falcon

Migrant breeding

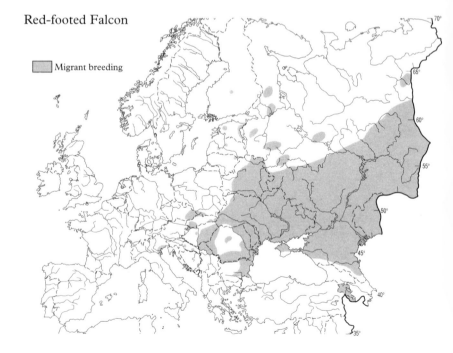

a narrow window of time. Large numbers are encountered on the Israeli coastal plain in early October. They migrate north farther west in Africa during spring migration; most birds enter Europe in southern Italy and Greece. Individuals tend to wander far afield during migration; they have been recorded as vagrants throughout most of the Western Palearctic.

Fine points

Amur Falcons have been recorded a few times in the Western Palearctic but, as its routes from north-eastern Asia to southern Africa and vice versa skirt the Middle East, it is to be expected as a vagrant. See under that species for distinctions.

Subspecies

None. However, sometimes considered conspecific with Amur Falcon *Falco amurensis*, but plumages are different for all age and sex classes. (See 'Fine points' in this account.)

Etymology

vespertinus is Latin for 'of the evening', probably referring to preferred hunting time.

Measurements

Length: 26–28 cm (27 cm)
Wingspread: 66–81 cm (73 cm)
Weight: 110–200 g (162 g)

Amur Falcon   *Falco amurensis*

(Eastern Red-footed Falcon)

Plate 40

Description

A slender, long-winged small falcon, a rare vagrant to the Western Palearctic. Adults have sexually dimorphic plumages; plumage of juveniles is subtly different from that of adult female. Plumage of adult male is similar to that of adult male Red-footed Falcon; those of adult female and juvenile are similar to that of juvenile Red-footed Falcon. Females are slightly larger than males. Iris is dark brown. Wingtips extend just beyond tail tip on perched birds.

Adult male. Head is dark grey, with paler grey cheeks and throat and darker grey moustache marks noticeable in good light. Body and upperwing coverts are dark grey. Flight feathers are silvery above and contrast with dark grey coverts; below they are dark grey and contrast with *white coverts*. Leg feathers and undertail coverts are bright rufous. Unbanded tail is dark grey. Cere, eye-rings, and legs are bright orange, with orange extending on to base of black beak.

Adult female. Crown and nape are dark grey and cheeks and throat are whitish, latter separated by short black moustache marks. White of cheeks extends on to sides of nape. Back, upperwing coverts, and flight feathers are blue-grey with black barring. Whitish underparts show narrow black streaking on breast, black spotting on belly, black barring on flanks, and pale rufous wash on unmarked leg feathers and undertail coverts. Underwings appear pale, with whitish coverts showing a variable amount of black markings, and dark secondaries and tips of primaries forming a wide dark band on trailing edges. Tail is blue-grey with black bands; subterminal band widest. Cere, eye-rings, and legs are orange, with orange extending on to base of black beak.

Juvenile. Similar to adult female, but back and upperwing coverts show rufous feather edges, black streaking on underparts is wider, leg feathers and undertail coverts are creamy, subterminal tail band is the same width as other dark bands, and flanks are not barred. Cere, eye-rings, and legs are yellow to yellow-orange, with colour of cere extending on to base of beak.

Second summer falcons

Second summer falcons (juveniles returning in spring) of both sexes resemble respective adults, as they have undergone a more or less complete body moult, not including coverts.

Second summer males usually appear somewhat like adult males; they have retained juvenile flight and tail feathers and show streaking on the breast. Central tail feathers have usually been replaced. First adult tail usually shows a dark subterminal band. Cere, eye-rings, and legs are bright orange like those of adult male.

Second winter males on autumn migration usually have some retained juvenile flight feathers, noticeable as 'patchwork' primaries.

Unusual plumages

No unusual plumages have been reported.

Similar species

Red-footed Falcon (Plate 40) adult males and juveniles appear similar to Amur Falcons. Adult male Red-footeds have dark underwing coverts and cheeks and do not show faint moustache marks. Juvenile Red-footeds have sandy-brown upperparts and reddish-brown (not black) streaking on underparts.

Eurasian Hobby (Plate 42) appears similar to juvenile Amur Falcon but is larger, has short pale superciliaries and shorter, unbanded uppertail, and uniformly coloured underwings. Adult Eurasian Hobbys also have rufous legs and undertail coverts like adult male Amurs but have streaked undersides, strong face pattern, and dark underwing coverts. Eurasian Hobbys often glide with wings held below level in a swift-like attitude.

Eleonora's Falcon (Plate 41) dark morph appears similar to adult male Amur but is larger with proportionally longer wings and tail and shows a pale line at base of flight feathers below. They also lack silvery uppersides of primaries, bright orange soft parts, white underwing coverts, and rufous leg and undertail feathers.

Sooty Falcon (Plate 42) appears similar to Amur Falcon but is larger. Juvenile Sootys have orangish-buff wash on underparts and lack pale banding on uppertail and on central feathers of undertail. Adult male Sootys are also overall grey like adult Amurs but lack silvery uppersides to primaries, white underwing coverts, and rufous leg feathers and undertail coverts.

Flight

Active flight is graceful, light and buoyant, with occasional quick wingbeats and longer periods of gliding, with constant adjustment of wings and tail. Soars on flat wings; glides with wrists below body and wingtips curved upwards. Hovers kestrel-like regularly but not persistently.

Moult

Adult moult is complete; it is begun in summer, suspended during migration, and completed on winter quarters. Juveniles replace many body feathers and central tail feathers in winter before return migration. First year birds begin flight feather moult in summer and complete in winter quarters.

Behaviour

Social and only occasionally encountered singly, then usually as vagrants. They prey on a wide variety of insects but feed young mostly on small mammals, frogs, lizards, and nestling and fledgling birds. When actively pursuing aerial insects, flight is somewhat erratic with jerky movements on fast wingbeats. They also perch on ground and run to capture prey.

Amur Falcons nest singly (not in colonies) and use unused nests of corvids. They form communal night roosts, except when breeding, and usually migrate in small to large flocks.

Status and distribution

A rare vagrant to the Western Palearctic, with a record from Turkey and two from Italy. It is likely that they occur more often but are overlooked in flocks of Red-footed Falcons.

Almost the entire population migrates to south-eastern Africa; this migration is thought to include a long sea crossing from the Indian subcontinent across the western Indian Ocean to southern Africa, usually bypassing the Western Palearctic. However, there are winter records from India.

Fine points
Amur Falcons are slightly smaller than the very similar Red-footed Falcon.

Subspecies
None. However, sometimes considered conspecific with Red-footed Falcon *Falco vespertinus*, but plumages are different for all age and sex classes.

Etymology
'Amur' and *amurensis* for Amuria or Amurland, the drainage area of the Amur River on the border between China and Russia, location of the first specimen.

Measurements
Length: 26–29 cm (27 cm)
Wingspread: 60–76 cm (69 cm)
Weight: 97–188 g (142 g)

Eleonora's Falcon *Falco eleonorae*

Plate 41; Photos, p. 352

Description

A gregarious, slender, long-winged, long-tailed large falcon that breeds only in the Western Palearctic in colonies on islands in the Mediterranean and off the north-west coast of Africa. It occurs in two colour morphs: light (~75 per cent) and dark (~25 per cent). However, the dark-morph falcons are of two types, not always separated under field conditions. All-dark (~2 per cent) have two dark genes, and darkish (~23 per cent) have one dominant dark gene and one light gene. Plumages of adult males and females are alike. Females are only slightly larger. Juvenile plumage is similar to that of adult. *Face pattern of small circular white cheek patches, lacking second short 'moustache marks' and pale nape areas, and two-toned underwings with darker coverts are distinctive.* Wingtips appear somewhat darker on uppersides of flying falcons and reach tail tip on perched falcons. Tail tip is usually wedge-shaped; central feathers are often elongated.

Light-morph adult. Head is dark slate to blackish-brown with whitish throat, and round cheek patches are separated by long narrow dark slate moustache marks. Back, upperwing and uppertail coverts, and uppertail are slate grey. Uppersides of primaries are blackish. Whitish breast is heavily streaked dark slate; sooty-rufous belly is less heavily streaked. Rufous leg feathers and undertail coverts are lightly marked with narrow slate streaks. *Underwings are two-toned*: dark blackish-brown coverts contrast with paler grey flight feathers. Pale undertail shows numerous faint narrow dark bands, subterminal wider. Cere and eye-ring are yellow on males. Female cere is bluish to whitish; female eye-ring varies from greenish-yellow to yellow. Short legs are yellow.

Light-morph juvenile is similar to adult, but crown, nape, and upperparts are dark brown with buffy feather edges. Buffy rounded cheeks and buffy throat are separated by narrow dark moustache marks. Blackish-brown flight feathers have wide buffy tips. Buffy underparts have narrow dark streaking and lack sooty-reddish cast of adults. Blackish-brown underwing coverts have heavy rufous markings and are noticeably darker than undersides of flight feathers; underwings are two-toned but with less contrast than those of adults. Uppertail shows faint pale banding on central feathers; undertail appears pale with numerous narrow dark bands, subterminal wider. Buffy undertail coverts are unmarked. Cere and eye-ring are bluish. Short legs vary from blue to pale yellow.

Second summer all birds show bands on undersides of flight feathers and upperside of tails (retained juvenile feathers); light-morph falcons lack sooty-reddish cast to undersides.

Dark-morph

Dark-morph birds are of two types: all-dark and darkish.

All-dark adult. Head, body, and wing coverts are dark slate to blackish, lacking any hint of paler cheeks or throat. Uppersides of flight feathers are blackish. Uppertail is grey. *Underwings are two-toned*: grey flight feathers contrast with black coverts. Undertail shows faint narrow dark banding. Sexual dimorphism as in light morph.

All-dark juvenile. Appears almost the same as all-dark adult, but most body feathers and coverts have dark rufous edges, and eye-ring and cere are bluish.

Darkish adults are similar to dark-morph adults but have whitish to dark dusky throats (sometimes white throat) and cheek patches separated by darker moustache marks and a variable amount of darker sooty-reddish cast or streaking on underparts.

Darkish juvenile is similar to light-morph juvenile but darker and with more heavily streaked underparts and heavily barred undertail coverts. Colour of soft parts as in light-morph juvenile.

Unusual plumages

Dilute-plumage and partial albino falcons have been reported from Italy.

Similar species

Lanner Falcon (Plate 44) juveniles appear quite similar to juvenile Eleonora's in silhouette, size, and underwing pattern but show dark forehead and pale crown and have belly streaking that ends abruptly on the lower belly.

Peregrines (Plate 46), particularly juveniles, can appear similar in flight, but have more compact and heavier bodies, wider wings, shorter tails, and lack two-toned underwings. Flight is more direct and powerful and less leisurely. Adults show white breast and barred belly.

Eurasian Hobbys (Plate 42), particularly juveniles, appear similar but are smaller, appear shorter tailed, have pale superciliaries and fewer tail bands, and lack two-toned underwings. Flight is also more swift-like. Adults have rufous restricted to leg feathers and undertail coverts.

Sooty Falcons (Plate 42) can appear somewhat similar. See under that species for distinctions.

Flight

Agile and swift powered flight is usually with slow, soft, deep wingbeats but is different when pursuing birds; then it is with rapid wingbeats of elastic wings, using entire wing, not just primaries. Soars on flat wings, sometimes with tail spread, sometimes with it folded, and often with small adjustments of wings and tail. Glides on flat wings, often giving single soft, short wing flap. They appear quite distinctive in flight because of their slow deep wingbeats.

Moult

Annual moult of adults is complete. Females begin during breeding, usually in September; males start later. Moult is suspended during migration and is completed on winter grounds. Post-juvenile moult is not well understood.

Behaviour

Gregarious aerial insect hunters that have adapted an unusual breeding strategy. They nest during the late summer and early autumn mainly on strategic islands and coasts in the Mediterranean Sea and off the Atlantic coast of north-western Africa and feed their young

on migrant birds that are captured over open water. They breed in colonies of up to hundred or more pairs. Adult males soar up above the colony and wait for incoming migrants. When one is sighted, they converge on it and begin attacking it with short stoops from above. Migrant birds are at times attacked by a dozen or more falcons in succession. Prey that are not eaten or fed to young are often cached for later retrieval.

Nests are placed on sandy ledges or in potholes on sea cliffs. Only the area close to the nest is defended by a pair. But the entire colony cooperatively defends the nesting area from other raptors and corvids. They are vocal during breeding but silent otherwise.

Entire population migrates to winter almost exclusively on Madagascar and the Mascarene Islands. Migrants are thought to travel eastward across the Mediterranean to the African–Syrian rift valley and then to turn south and travel along the Red Sea and African coast to Madagascar, but an inland record from Mali would seem to indicate a direct trans-African route for some falcons. They feed almost exclusively on flying insects while migrating and on the winter grounds. Birds returning to the breeding areas also feed primarily on insects until breeding commences.

Insects are hunted aerially much in the manner of Eurasian Hobbys, and those captured are eaten by the falcons in flight.

Status and distribution

Breed exclusively in the Western Palearctic in very local colonies on islands and coastal cliffs in the Mediterranean Sea and along the north-west Atlantic coast of Africa. The entire population migrates in autumn to main winter quarters on Madagascar and the islands of Reunion, Mauritius, and Rodriguez, but some apparently winter in Tanzania, perhaps also other areas of south-eastern Africa; however, these reports are unsubstantiated.

Several unsubstantiated winter records exist for the Mediterranean. Reliable reports exist of regular vagrants throughout Europe.

Eleonora's Falcon

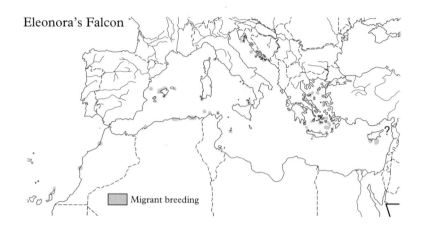

Migrant breeding

Fine points

The soft, slow, deep wingbeats are distinctly different from those of any other large falcon.

Subspecies

Monotypic.

Etymology

Named for Princess Eleonora of Arborea (*ca.* 1350–1403) of Sardinia, because of her protection of birds of prey.

Measurements

Length: 35–44 cm (40 cm)
Wingspread: 84–103 cm (94 cm)
Weight: 300–460 g (370 g)

Sooty Falcon *Falco concolor*

Plate 42; Photos, p. 354

Description

A medium-sized, long- and narrow-winged, summer inhabitant of desert islands and deserts of north-east Africa, Sinai, Israel, and Arabia. Adult plumage is similar for the sexes; females are slightly larger. Juvenile plumage is different from that of adults. First summer birds are recognizable by transition in plumage. Contrary to earlier accounts, there is no dark colour morph of this falcon. Central tail feathers are elongated in all plumages; tail tip on flying falcons appears wedge-shaped. On perched birds, wingtips extend noticeably beyond tail tip.

Adult male. Head, body, and wing and tail coverts are sooty grey. Some males have a small whitish throat patch and faint darker moustache marks. Primaries are black above and contrast with coverts. Underwing is paler than body, with primary tips black. Unbanded tail is darker sooty grey; in good light darker band on tail tip is sometimes noticeable. Cere and eye-rings are bright yellow to orangish-yellow; colour extends on to base of black beak. Legs are orange-yellow to orange.

Adult female is like male but overall coloration averages darker sooty grey, particularly on head, with less contrast between primaries and coverts, but with overlap in this character. Uppertail coverts and rump are paler sooty grey. Cere and eye-rings are lemon-yellow; colour extends on to base of black beak. Legs are yellow to orangish-yellow, usually duller than those of adult male.

First summer birds. One-year-old falcons returning from winter areas have undergone a more or less complete moult. General appearance is like that of adults, but many birds appear quite mottled due to retained juvenile body feathers. All usually show white throat. New tail usually has barely noticeable narrow pale bands on inner feather webs, visible only from below. Cere, eye-rings, and legs are yellow.

Juveniles (sexes alike). Head is covered by brownish-grey 'hood', with narrow buffy feather edges and extensions below eyes forming moustache marks and similar shorter ones behind ear coverts; these marks separate cinnamon-buff throat, cheeks, and sides of nape. Back and upperwing and uppertail coverts are brownish-grey with narrow buffy feather edges. Primaries above are dark brown. Underparts are orange-buff with brownish-grey centres to feathers that are larger on the upper breast but narrower on belly, thus appearing as streaks. Seen from a distance, a darker band across breast is usually noticeable. Tail is dark brown; *central pair of feathers are unbanded*, and others have four or more narrow white bands on inner webs that are visible only on undertail, but with wide tip unbanded. Uppertail shows no banding, only narrow pale band across tips. Cere and eye-rings are blue. Beak is two-toned; pale yellow base and otherwise black. Legs are pale yellow to yellow.

Unusual plumages
No unusual plumages have been reported.

Similar species
During the summer, there are no other medium-sized falcons in the breeding area of the Sooty Falcon. Confusion could be possible, however, with grey-backed adult Barbary Falcon. However, during migration, at least three other similar falcons migrate through their breeding range.

Red-footed Falcons (Plate 40), particularly juveniles, are quite similar to juvenile Sootys in silhouette and pattern. But Red-footeds are smaller, show pale banding on uppertail and on central feathers of undertail, sandy, not grey-brown, back and upperwings, and pale collar that extends across nape. Perched Red-footeds show pale forehead and crown that contrast somewhat with darker areas around eyes. Adult male Red-footeds are also overall grey like adult Sootys but are more blue-grey and have silvery uppersides to primaries and rufous leg feathers and undertail coverts.

Eurasian Hobby (Plate 42) juveniles are similar to juvenile Sootys in size, silhouette, and pattern but have relatively shorter wings and thicker, more well defined streaking on buffy, not orange-buff, underparts, and lack dark breast band and two-toned pattern on underwings. Adult Hobbys have rufous leg feathers and undertail coverts.

Eleonora's Falcons (Plate 41) are similar in silhouette, with dark-morph falcons resembling adult Sootys and light-morph ones resembling juveniles, but have relatively longer tails, fly with slow, deliberate wingbeats, and have two-toned underwings—coverts darker than flight feathers. Dark-morph Eleonora's are darker, more blackish than are adult Sootys. Some may have white throats like some Sootys. Light-morph Eleonora's plumage is similar to juvenile Sootys, but they have a dark rusty cast to undersides, have smaller more distinct white cheek patches, and lack pale nape patches and second 'moustache marks' behind ear coverts.

Barbary Falcon (Plate 46) adults also have grey backs, but colour is more blue-grey. They are also much larger, more compact, and shorter tailed. They have noticeable dark moustache marks and pale undersides with barring on belly.

Flight

Active flight is with fast, stiff wingbeats. Soars and glides on flat wings.

Moult

Annual moult of adults is complete but is accomplished completely on wintering grounds from November to April. Post-juvenile moult, including flight feathers are rectrices, begins during first winter but is often incomplete.

Behaviour

Slender, swift, aggressive aerial hunters that are mainly insectivorous during winter but feed mostly on birds in desert breeding areas. Breeding is delayed until late summer to take advantage of numbers of autumn bird migrants to feed young. Hunting is usually by one of two methods; searching from an exposed perch, with a direct rapid flight toward a bird spotted flying over the desert or sea. Then the bird is pursued relentlessly until it escapes or is captured. Sootys also soar up to great heights and attack migrants from above with the same persistency. In island colonies, as many as 10 or more Sootys take turns attacking the same bird. In the desert they also hunt by gliding along a wadi trying to flush birds from small bushes. Sootys' activity periods vary according to prey availability; they are often crepuscular.

Insects are usually eaten on the wing, whereas birds are taken to perches to pluck and eat. A few small mammals, bats, and even fish have been reported in their diet. But the vast majority of summer food is birds, many of which are agile flyers like swallows, swifts, and bee-eaters. Some are surprisingly large, ranging up to the same size or even larger than the Sooty.

Nest is usually on a cliff ledge or pothole in the desert or desert island, but some have been found under bushes and in an old Reef Heron nest on small islands, and even under a camel marker of sandstone slabs in mid desert. All nests are completely shaded. Sooty Falcons are somewhat vocal when breeding but are silent otherwise. They are aggressive to any bird of prey or corvid that comes into the nesting area.

Sootys are somewhat gregarious during winter and feed mainly on flying insects.

Status and distribution

An uncommon and local summer breeder in deserts of Libya, Egypt, Sinai, southern Israel, and desert islands of the northern Red Sea. They also occur on desert islands around the Arabian peninsula.

They are most probably more numerous than the paucity of breeding records would indicate; intensive efforts in Israel have raised number of breeding pairs from two to almost 100 in less than 10 years. Also they are rather common on winter quarters.

All falcons leave the breeding areas; most spend the northern winter on Madagascar, but some also winter along the south-eastern coastal plain of Africa.

Vagrant records exist for Cyprus, Malta, and Spain.

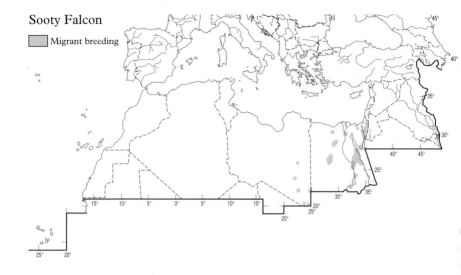

Sooty Falcon

Migrant breeding

Fine points
See Frumkin and Clark (1988) for refutation of a dark colour morph.

Subspecies
None.

Etymology
concolor is Latin for 'same colour', most likely referring to its uniform coloration in adult plumage.

Measurements
Length: 29–33 cm (31 cm)
Wingspread: 77–87 cm (82 cm)
Weight: 225–280 g (255 g)

Merlin   *Falco columbarius*

Plate 43; Photos, p. 358

Description

The dashing, short-winged Merlin, the Western Palearctic's smallest falcon, breeds across northern Europe. *It lacks the bold moustache mark of other falcons*, having at most only a faint one. Sexes differ in adult plumage and size, with females being noticeably larger. Juvenile plumage is similar to that of adult female. *Wingtips fall quite short of tail tip on perched birds*. Iris is dark brown. Cere and eye-ring are greenish-yellow to yellow. Legs are orangish-yellow. (The following descriptions apply to widespread race *aesalon*.)

Adult male. Head consists of blue-grey crown and nape that are marked with fine black shaft streaks, short narrow pale superciliary line, pale area on cheeks bordered by faint blue-grey areas, the front ones forming faint moustache marks below each eye, and a pale unmarked throat. Rufous-buffy markings form pale collar across hind-neck. Back and uppertail and upperwing coverts are blue-grey, with narrow black shaft streaks and grading paler toward rump and uppertail coverts. Uppersides of flight feathers are black. Uppertail is blue-grey with wide black subterminal band and narrow blue-grey terminal band; some tails may also have narrow incomplete black banding. Underparts, leg feathers, and undertail coverts are rufous-buff with narrow black streaking; colour darker on leg feathers. Underwing is uniformly marked and appears somewhat dark on distant birds.

Adult female. Head consists of brown crown and nape that are marked with fine black shaft streaks, short narrow buffy superciliary line, buffy area on cheeks bordered by brown areas, the front ones forming faint moustache marks below each eye, and a whitish unmarked throat. Buffy markings form pale collar across hind-neck. Back and upperwing coverts are dark brown with narrow black shaft streaks, some buffy cross-barring, and often a faint greyish cast (older birds?). Uppertail coverts are also brown but always with a greyish cast. Uppersides of flight feathers are brown with some rufous-buffy spotting. Uppertail is brown with rufous-buffy to whitish bands, narrow pale band on tail tip, and sometimes a greyish cast. Underparts are buffy with dark brown streaking, becoming thicker and blob-like on belly and tending to barring on flanks. Leg feathers and undertail coverts are buffy with short fine dark brown streaks. Underwing is uniformly marked and appears somewhat dark on distant birds.

Juvenile. Like adult female, but back is slightly darker brown and females lack greyish cast on rump and uppertail coverts.

Other races. Merlins of other races that occur in the Western Palearctic are similar in overall pattern but differ in size and in intensity of coloration. Falcons from Iceland, *F. c. subaesalon*, are larger and longer-winged and adult females and juveniles average somewhat darker, particularly lacking rufous feather edges on upper backs. Adult males from Iceland are indistinguishable from nominate males. Some Merlins wintering in Turkey and the Middle East are *F. c. pallidus*; they are much paler; backs of males are pale grey and females and juveniles appear more rufous than brown. Markings on underparts are less intense. Vagrant *F. c. columbarius* from North America are similar in size but are much darker; adult males have black tails with narrow grey bands.

Unusual plumages

Records and specimens exist of partial albinos and albinos. A dilute-plumage specimen was collected in Sweden.

Similar species

Eurasian Hobbys (Plate 42) can appear similar. See under that species for differences.

Red-footed Falcons (Plate 40) can appear similar but are larger with longer wings. Wingtips reach tail tip on perched Red-foots. Juvenile Red-footeds have noticeable white cheeks and napes and dark moustache marks. Adult male Red-footeds are also grey but head is completely grey, upsides of primaries are silvery, and leg feathers and undertail coverts are rufous.

Eurasian Sparrowhawks (Plate 23), particularly grey-backed adult males, can appear similar in ground-hugging flight but have wider wings, rounded wingtips, and longer tails.

Peregrines (Plate 46) in a soar can appear similar, but are much larger, have relatively longer wings, distinct moustache marks, and larger heads.

Both Kestrels (Plates 38, 39) are larger, have longer wings and tail, and show contrasting two-toned pattern on uppersides in flight.

Flight

Active flight is direct with strong, quick wingbeats. Soars on flat wings with tail somewhat fanned. Glides on flat wings or with wrists lower than body and wingtips curved upwards. Usually does not hover.

Merlins sometimes use an unusual flight mode when attacking a bird from a low altitude; they flap their semi-closed wings in quick bursts, interspersed with glides. The resulting flight is rapid and undulating, similar to that of many passerines. This may allow them to be mistaken for a 'harmless' bird until it is too late for escape.

Moult

Annual moult of adults is complete, beginning in early summer and completed by end of year, with females moulting earlier than males. Post-juvenile moult begins in spring with body moult but may include central tail feathers. Completed by end of year. Subsequent moults are like those of adult.

Behaviour

Dashing falcons that hunt birds on the wing. Their rapid hunting flight is typically direct over open forest, grassland, or beach, using their speed to surprise a flock of birds and then to snatch one that is slow to react in mid-air. They also hunt from exposed perches, making rapid forays after prey. Merlins fly fast enough to catch swallows, swifts, and waders and are capable of sudden changes in direction. They are capable of making spectacular aerial manoeuvres. Small birds are often forced to fly up to great altitudes, and are captured when they try to drop back to earth. As in many other falcons, pairs have been reported hunting birds cooperatively. Fledglings regularly capture insects in flight and eat them on the wing. Mammalian prey is also taken, particularly by fledglings and in times of lemming abundance.

Courtships flights are usually inconspicuous and often consist of males flying around proposed nest site with fluttery wing beats and vocalizing. Nests are placed on ground (most often in Britain and Iceland) or in old tree nest, usually corvid (most often in Scandinavia, Finland, and Russia), but a few also on cliff ledges (Scotland). Their primary vocalization is a rapid, high-pitched call, similar to that of the Eurasian Hobby. Merlins regularly go out of their way to harass larger birds, including gulls and other raptors.

They form communal night roosts in winter, most often with fewer than 10 birds, but sometimes up to 60 or 70.

Status and distribution

Summer inhabitants of central and northern Scandinavia, Finland, and northern Russia; they move south into central and southern Europe for winter. Also resident in Britain and Ireland, except for southern England, and summer inhabitants in Iceland. Most, but not all, of the latter move into the British Isles for winter. Central Asian Merlins winter locally in Turkey and the Middle East. A vagrant *columbarius* from North America was reported from Iceland.

Fine points

Icelandic Merlins are almost impossible to distinguish in the field.

Subspecies

Falco c. aesalon occurs throughout northern Western Palearctic from Ireland and Britain through Scandinavia and northern Russia. *F. c. subaesalon* breeds in Iceland. *F. c. pallidus* winters in Turkey and Middle East. *F. c. columbarius* from North America have been recorded as a vagrant in Iceland and twice as vagrants to Britain, but the latter records are in question.

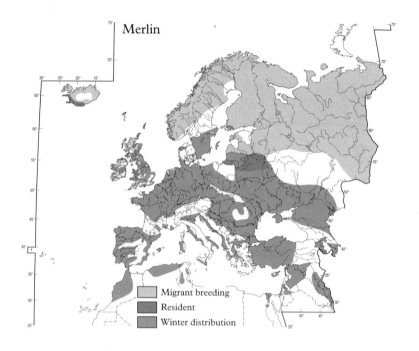

Etymology

columbarius is Latin for 'pertaining to a dove (pigeon)', referring to Merlin's original North American name of 'Pigeon Hawk', for its resemblance in flight to the Pigeon. Merlin derives from the Old French *esmerrilon*, the name for this species.

Measurements

Length: 25–31 cm (28 cm)
Wingspread: 54–69 cm (62 cm)
Weight: male 150–190 g (175 g)
          female 180–245 g (220 g)

Eurasian Hobby *Falco subbuteo*

Plate 42; Photos, p. 356

Description

A widespread, medium-sized, swift, and elegant falcon with long, narrow wings and a relatively short tail. *Flight silhouette appears sickle-like* due to curved leading edge of outer wings, and in flight they often appear quite swift-like. Wing proportions are Peregrine-like, with short arms and long hands. Sexes are alike in plumage; females average slightly larger. Juvenile plumage is similar to that of adults. Iris is dark brown. On perched birds, wingtips extend just beyond tail tip.

Adult. Head is composed of grey crown, often with narrow rufous-buffy feather edges, short narrow pale superciliary, creamy unmarked cheeks and throat separated by long narrow black moustache marks, and creamy unmarked areas on sides of nape partially separated from cheeks by second set of short black 'moustache marks'. Back and upperwing and uppertail coverts are dark blue-grey; falcons in worn plumage, especially females, appear somewhat dark brown on upperparts. Uppersides of flight feathers are blackish. Whitish to buffy underparts have thick, well-defined black streaking; *individual streaks often extend from throat to lower belly.* Because of heavy markings, underparts can appear uniformly dark on distant falcons, with contrasting white throat and cheek noticeable. Uniformly marked underwings appear dark. Leg feathers and under-tail coverts are rufous. Upperside of folded tail appears unbanded with a blue-grey cast (central feathers), but underside appears banded, with eight or more sets of even-width dark brown and rufous-buffy bands. Central tail feathers are often elongated. Cere and eye-ring are yellow; legs are orange-yellow.

Juvenile is similar to adult, but upperparts are dark brown, showing buffy feather edges in fresh plumage. Underparts are heavily marked with wide dark brown streaks; *individual streaks often extend from throat to lower belly.* Leg feathers and undertail coverts are creamy, with narrow dark streaks. Folded tail appears unbanded dark brown on upperside, with a wide pale terminal band, but on undersides shows more distinct even-width dark brown and buffy bands. Cere, eye-rings, and legs are yellowish-green, later becoming yellow.

Unusual plumages

An adult specimen from Russia showed two all-white primary upperwing coverts. A dilute-plumage specimen from Ethiopia had pale whitish-grey tail and flight feathers and underparts, but normally coloured upperparts.

Similar species

Red-footed Falcon (Plate 40) adult males and juveniles appear similar in silhouette and plumage. See under that species for differences.

Eleonora's Falcon (Plate 41) light-morph birds appear similar. See under that species for differences.

Merlins (Plate 43) are smaller, have shorter wings and banded uppertails, and lack whitish cheeks and well-defined moustache marks. Wingtips fall quite short of tail tip on perched birds. Adult male also has blue-grey upperparts but has wide black terminal band on tail and lacks rufous undertail coverts.

Peregrine (Plate 46) juveniles appear similar in flight but are larger and have wider bases of wings and thicker single moustache mark. Their flight is stronger and bolder. Wingtips just fall short of tail tips on perched juvenile Peregrines.

Flight

Powered flight when hunting is rapid with fast, stiff wingbeats, but they also have a light, buoyant, more leisurely flight. Soars on flat wings, usually with wingtips pulled back and wings bent at wrist, and with tail somewhat fanned. Glides often swift-like with wrists lower than body and wingtips pointed downwards below the horizontal. Stoops after prey in rapid dives, when it appears 'tear drop' in shape. Hovers occasionally for a short while, usually when prey has gone to cover.

Moult

Annual moult of adults is complete; it begins in late summer, is suspended during autumn migration, and is completed on winter grounds. Juveniles begin body moult in early spring while still on winter quarters.

Behaviour

Graceful, elegant, and acrobatic falcons that capture insects and birds on the wing. They hunt birds with Peregrine-like lightning stoops, sometimes ending with a short upward swoop. They hunt insects in a less dramatic, more leisurely fashion, again often ending with a short upsweep; insects are usually eaten on the wing. They also eat bats, small mammals, and reptiles and are often active at dawn and dusk. On occasion, they resort to piracy, particularly robbing Common Kestrels of their prey.

In winter and on migration, they are social, forming feeding flocks and communal night roosts, often in association with other insect-eating small falcons.

Hobbys nest in trees, most often using old corvid nests. They do not begin nesting until summer, most likely waiting until corvid nests are available but also because more passerine fledglings are available then to feed their young.

Male courtship flights are feats of aerobatics, usually accompanied by vocalizations. Pairs investigate many areas for possible nests, again accompanied by vocalizations. Hobbys are usually silent outside of the nesting season.

Status and distribution

Uncommon to common summer breeders throughout the Western Palearctic almost everywhere there are trees and corvids but avoiding dense continuous forests. Almost all birds migrate to southern Africa for the winter, but a few winter records exist. Usually seen in small numbers at raptor migration concentrations.

Eurasian Hobby

Migrant breeding

Fine points

Juvenile's tail has wide buffy terminal band; that of adult is narrow.
Tip of folded tail often appears wedge-shaped.

Subspecies

The Western Palearctic race is *Falco s. subbuteo*.

Etymology

'Hobby' comes from the Old French *hobe*, meaning 'to jump about', for its agility in capturing insects on the wing. *sub* and *buteo* are Latin for 'somewhat' and 'a kind of hawk or falcon'.

Measurements

Length: 29–33 cm (31 cm)
Wingspread: 74–83 cm (78 cm)
Weight: male 131–223 g (193 g)
        female 141–325 g (237 g)

Lanner Falcon   *Falco biarmicus*

Plate 44; Photos, p. 360

Description

A large slender falcon, a rare local resident in drier, hilly areas of central and eastern Mediterranean countries but more common, yet still local, in arid parts of northern Africa. Falcons of two north African subspecies are identical to each other but differ from European Lanners. Sexes are almost alike in plumage; females are on average darker and noticeably larger. Juvenile plumage is different from that of adult; second winter plumage is recognizable. Iris is dark brown. Beak is yellowish at base and black on tip. Following account describes European race *feldeggii*. Wingtips fall somewhat short of tail tip on perched falcons.

Adult. Head pattern consists of dark grey forehead and either dark grey (♀) or buffy-rufous (♂) crown, buffy-rufous (can be faded to buffy) nape, narrow black eyelines behind each eye that extend on to nape, and whitish cheeks and throat separated by long, narrow black moustache marks below each eye. Ashy-grey to brownish ashy-grey back feathers and upperwing coverts have a variable amount of narrow pale grey to buffy barring and feather edging. Rump and uppertail coverts are brighter blue-grey. Ashy-grey tail has many narrow pale grey to buffy bands. Markings on creamy underparts are narrow dark streaks on breast, spots on belly, and barring on flanks, undertail coverts, and leg feathers. Underwings usually appear somewhat pale except for dark tips of outer primaries; however, greater underwing coverts are heavily marked on some birds, resulting in a darkish band through each underwing. Cere, eye-rings, and legs are yellow to orange-yellow.

Second winter Lanners are similar to adults but feathers of upper back are darker and lack pale cross-barring and feather edges, rufous on nape (and crown when rufous) has many narrow black streaks, and underparts are more heavily marked with large spots on breast and larger spotting and wider barring on belly and flanks.

Juvenile. Head consists of buffy to dark brown crown, whitish markings on nape, white cheeks, and long narrow black moustache marks. Back and upperwing coverts are uniformly dark brown, as are uppersides of flight feathers. Dark brown tail has numerous incomplete rufous-buffy bands, with central pair often unbanded; tail tip shows wide pale terminal band. Some juvenile Lanners have pale Saker-like ovals or spots in tails. Whitish breast and upper belly are heavily streaked dark brown; some birds appear almost solid dark below. Whitish lower belly, leg feathers, and undertail coverts are only lightly streaked, with noticeable line of contrast with darker upper belly. Underwing coverts are heavily marked and contrast somewhat with paler, somewhat greyish, flight feathers; bases of primaries are especially paler. Wingtips fall short of tail tip by 2–6 cm on perched juveniles. Cere and eye-ring are blue-grey; legs are yellow.

North Africa Lanner Falcons

North Africa Lanner Falcons of races *erlangeri* and *tanypterus* are indistinguishable from each other (latter average slightly larger) but are overall paler in colour than European ones.

Adults have creamy to rufous crowns (sexes alike), with short black streaks often forming a black mark across forecrown and over each eye. Narrow black eye-lines extend on to nape. Back and upperwing coverts are on average paler, with more pale feather edges and cross-barring. Tail is more blue-grey. Underparts are more lightly marked and lack barring on flanks. Wingtips reach tail tip on perched adults.

Second winter falcons are overall darker-backed than are adults, with narrow dark streaking on rufous crowns and napes, and are more heavily spotted on underparts.

Juveniles are similar to juvenile *feldeggii*; main difference is pale crown, often giving a white-headed appearance. Wingtips fall a bit short of tail tip on juveniles.

Unusual plumages

No unusual plumages have been reported.

Similar species

Saker Falcons (Plate 45) appear similar to juvenile Lanners in colour and pattern but are larger and heavier, with streaking on uniformly coloured crowns, paler undersides of flight feathers, particularly primaries, kestrel-like two-toned pattern on upperwings, pale superciliary lines that extend on to nape, and, when perched, wingtips that fall quite short of tail tip. They also lack Lanner's dark mark on forecrown (not always obvious on dark-headed Lanners) and juvenile Lanner's sharp line of contrast between heavily streaked upper and lightly streaked lower belly. (Adult Lanners are distinguished by grey backs and rufous crowns and napes.)

Peregrines (Plate 46) are similar but have thicker moustache marks, dark crowns and napes, relatively shorter tails, and darker, uniformly coloured underwings. Adult Peregrines' underparts are marked with short bars, not spots, and their tails appear more two-toned on distant Falcons. Peregrines in flight appear much different from Lanners; flight is more vigorous, with more stocky build obvious.

Barbary Falcon (Plate 46) adults when perched appear similar to north African adult Lanners; both have blue-grey upperparts and rufous napes and wingtips that reach tail tip. But adult Barbary head pattern shows dark crown, with rufous restricted to nape, and wider, less well defined black lines behind eyes. Their underparts are more orangish-buff, not whitish, and are marked on belly and flanks with short bars, not spots. Barbary Falcons in flight appear much different from Lanners; flight is more vigorous, with more stocky build obvious.

Eleonora's Falcon (Plate 41) juveniles appear similar to juvenile Lanners. See under that species for distinctions.

Flight

Powered flight is with slow, measured, shallow wingbeats and appears somewhat kestrel-like. Flight is more buoyant than that of other large falcons due to lighter wing loading but appears rapid when actively pursuing prey, nevertheless lacking the power of a Peregrine. Soars and glides on level wings, but sometimes glides with wrists below body and wingtips swept up. Soaring silhouette with forewings forming straight line is distinctive. Lanners hover occasionally, but more often hang kite-like on an updraft from a cliff face.

Moult

Annual moult of adults is complete; it begins in spring and is completed by late autumn. Juveniles begin a body moult in winter; they moult flight and tail feathers at the same time as adults do.

Behaviour

Prey mainly on birds, but diet is extremely varied, as they are generalists exploiting a variety of food resources. They take many small- to medium-sized mammals including bats, as well as reptiles, snakes, and insects. They also have been observed pirating food from smaller raptors and Black Kites and eating carrion. They employ a variety of hunting techniques: perch hunting, terrain following flight, soaring, and walking on the ground grabbing insects. Prey is taken equally on ground or in the air.

Lanners are mainly solitary birds, but often small groups of non-breeding birds will congregate at food concentrations, particularly in Africa south of the Sahara. Adult pairs regularly hunt as a coordinated team, particularly prior to breeding and with fledglings.

They nest usually on cliffs or crags but will use the stick nest of other raptors in trees. Display flight includes mutual high soaring, undulating flight, and flight aerobatics, as well as spectacular plunges by males.

Lanners are vocal only during the breeding season. Vocalizations are like those of other large falcons.

Status and distribution

Rare, local residents of dry open hilly lowlands of Italy, the former Yugoslavia, Greece, Turkey, Israel, Jordan, Sinai, and Egypt along the Nile River. More common but still local in various arid habitats, including deserts, in Morocco, Algeria, Tunisia, and Libya. An old breeding record exists for Spain.

Lanner Falcon

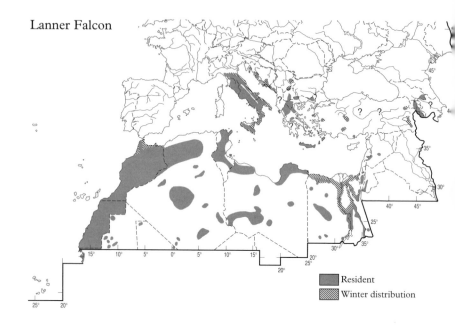

Resident

Winter distribution

Juveniles disperse, sometimes to considerable distances. Vagrants are reported from various parts of southern Europe.

Fine points

Blackish markings on the underwing coverts of adults are vertical; those of very similar adult Barbary Falcons are horizontal.

Subspecies

F. b. erlangeri occurs in north-western Africa, *F. b. tanypterus* occurs from north-eastern Africa into the Middle East, and *F. b. feldeggii* occurs from Italy eastward through Turkey. There are no differences in plumage between *erlangeri* and *tanypterus*.

Etymology

'Lanner' is apparently derived from the Old French *lanier*, for 'cowardly', presumably given to it by Middle Age falconers for its less aggressive demeanour compared to that of the Peregrine. Another interpretation is that 'Lanner' comes from the Italian *lanier* for 'wool-like'. *biarmicus* comes from Latin *bi* and *armicus* for 'two' and 'armed', apparently in reference to bill's tip and 'tooth'.

Measurements

Length: 38–49 cm (44 cm)
Wingspread: 90–115 cm (103 cm)
Weight: male 450–650 g (550 g)
　　　　 female 550–800 g (650 g)

Saker Falcon   *Falco cherrug*

Plate 45; Photos, p. 362

Description

A long-tailed, large brown falcon, a rare inhabitant of the steppes of eastern Europe and Turkey. It is larger than other large falcons likely to be encountered at same time and place; only juvenile Peregrines and Lanner Falcons are also brownish. Underwing is usually two-toned: darkly marked coverts contrast with paler flight feathers. Sexes are alike in plumage; females are noticeably larger. Juvenile plumage is almost identical to that of adult. Wingtips fall 4–8 cm short of tail tip on perched birds. Iris is dark brown. Beak is pale at base and black on tip.

Adult. Pale head is composed of unmarked throat and forehead, light to heavy brown streaking on crown and nape, paler superciliary lines, dusky eyelines behind each eye, faintly streaked cheeks, and narrow dusky moustache marks below each eye. Dark brown back feathers and upperwing and uppertail coverts have narrow pale to rufous feather edges. Flight feathers above are darker brown and usually contrast with paler tawny-brown wing coverts; *most adults show kestrel-like two-toned pattern on upperwings.* Brown tail has many narrow buffy to rufous-buffy bands, with central pair sometimes unbanded; oval-shaped incomplete bands *appear as large pale spots on tails of many falcons.* Whitish underparts are lightly to heavily marked with brown spots, appearing sometimes as streaks on belly and flanks. Lightly marked adults show short narrow streaks on breast. Undertail coverts are unmarked. Undersides of wings are pale except for dark tips of outer primaries and sometimes dark greater underwing coverts; latter shows as darkish band through underwing. Cere, eye-rings, and legs are yellow.

Juvenile is similar to adult, but underparts are more heavily streaked. Underwing coverts are usually more heavily marked, therefore darker, and contrast more strongly with pale flight feathers. Brown tail feathers most often *show pale ovals or spots* on inner webs and often also on outer webs, but some birds show bands. Cere, eye-rings, and legs are blue-grey.

Note: Intensity of brown coloration is variable. Darker birds show dark heads, heavier brown streaking overall, and narrower pale feather edges on back and upperwing coverts, and paler birds appear quite pale-headed, with narrow streaking on underparts and wide pale feather edges on back and upperwing coverts.

Unusual plumages

Dilute-plumage birds are *café-au-lait* coloured and are favoured by Arab falconers, who also have partial albinos and other differently col-oured falcons from Central Asia. (See 'Fine points' in this account.)

Similar species

Lanner Falcons (Plate 44) appear similar. See under that species for distinctions.

Peregrines (Plate 46) are similar but are smaller, with shorter tail, narrower wing bases, thicker moustache marks, and darker, uni-formly coloured underwings. Wingtips reach, or almost reach, tail tip on perched birds. Adult Peregrines have dark blue-grey backs, clear unmarked breasts, and two-toned uppertails.

Gyrfalcons' (Plate 45) range hardly overlaps with that of Saker. They are larger, heavier bodied, with relatively smaller heads and noticeably wider tails, and lack any rufous brown coloration.

Flight

Powered flight is with slow, powerful, somewhat elastic wingbeats. Resulting flight is rapid. Soars and glides on level wings, but some-times glides with wrists below body and wingtips swept up. Sakers hover briefly on occasion like a large kestrel, particularly when look-ing for prey that has taken cover.

Moult

Annual moult of adults is complete; it begins in spring and is com-pleted by late autumn. Juveniles begin a body moult in winter; they moult flight and tail feathers at the same time as adults do.

Behaviour

Prey mainly on small mammals, especially susliks and other microtines, but also take a wide variety of birds, particularly ground nesters. Their usual hunting methods are to search from an elevated perch commanding a good view of terrain and launch out after prey when sighted and to fly rapidly in a terrain hugging manner and surprise prey after coming over a hillock. Less often they hunt from a high soar and stoop rapidly when prey is sighted; they have been observed walking on ground capturing insects. Most prey is taken on ground but can be taken in air. Sakers possess much stamina and will tail chase until prey is exhausted.

They usually nest in trees using old stick nest of other raptors or Raven, but sometimes use a stick nest on a cliff or crag. Display flights include mutual high soaring and flight aerobatics, but apparently do not include undulating flights.

Sakers are usually solitary, and are vocal only during the breeding season. Vocalizations are like those of other large falcons.

Status and distribution

Rare summer residents of steppe and wooded steppe of east central Europe (mainly Russia), with range extending west locally to the Ukraine, the former Czechoslovakia, Hungary, Romania, Bulgaria, the former Yugoslavia, central Turkey, and, formerly, Austria. Most birds leave the breeding areas for winter, particularly those in Russia; however, some may remain in western parts. They winter locally on Mediterranean coasts and in the Middle East and lower Nile River drainage. Small numbers are recorded on migration in the eastern Mediterranean.

Fine points

Asian adult Sakers (*F. c. milvipes*) appear different; their backs are marked with rufous barring and their tails have rufous bands. Juveniles are like European birds. As falcons escape their Arab owners, these, as well as dark and whitish Falcons could be seen in the Middle East.

Saker Falcon

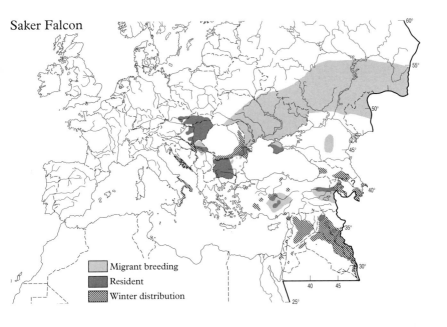

Migrant breeding
Resident
Winter distribution

Subspecies

F. c. cherrug occurs in eastern Europe.

Etymology

'Saker' after the bedouin tribe 'Bani Sakr' of southern Jordan, who specialized in capturing this species for falconry; *cherrug* is from the Hindi *charg*, the name for the female.

Measurements

Length: 43–60 cm (51 cm)
Wingspread: 105–128 cm (115 cm)
Weight: male 665–900 g (775 g)
      female 900–1300 g (1050 g)

Gyrfalcon   *Falco rusticolus*

(Gyr Falcon)

Plate 45; Photos, p. 364

Description

The Western Palearctic's largest falcon, a rare inhabitant of arctic regions of Iceland, Norway, Sweden, Finland, and Russia. They are *heavier bodied and broader winged* than are other falcons, with *wide, noticeably tapered tail. Their heads appear relatively small compared to their bodies.* Underwing is usually two-toned: heavily marked coverts contrast with paler grey flight feathers. Sexes are alike in plumage; females are noticeably larger. Juvenile plumage is different from that of adult. Wingtips reach only three-quarters of way down tail on perched birds. Iris is dark brown. Beak is pale at base and black on tip.

Adult. Head is slate-grey with pale superciliary lines, narrow black moustache marks below each eye, and white unmarked throat, and often with pale streaking on cheeks, crown, and nape. On paler falcons cheeks appear whitish with narrow slate streaking. Slate-grey back and upperwing coverts are crossed with paler grey barring. Rump and uppertail coverts are grey with slate barring and appear paler than back and upperwing coverts. Uppersides of flight feathers are dark slate-grey with pale barring. Tail is either grey with slate-grey bands or slate-grey with grey bands. Whitish underparts are marked with short, narrow dark streaks and spots on breast, larger dark spots on belly, and dark barring on flanks, leg feathers, and undertail coverts. Cere, eye-rings, and legs are yellow.

Juvenile. Head is slate-brown with pale superciliary lines, narrow black moustache marks below each eye, white throat, and pale streaking on cheeks and nape. Slate-brown back and upperwing and uppertail coverts have narrow pale feather edges, with some pale spotting on tertials, upperwing and uppertail coverts, and lower scapulars. Uppersides of flight feathers are slate-brown. Grey-brown tail shows buffy narrow, sometimes incomplete bands. Creamy underparts are heavily marked with wide dark brown streaks. Cere, eye-rings, and legs are blue-grey. *Note*: There is variation in the amount and intensity of dark coloration in both age plumages, with some individuals being quite pale, others somewhat dark.

Juvenile white-morph Falcons occur as vagrants, most probably from Greenland. Head is white with narrow brown shaft streaking on crown and nape, blackish line behind each eye, and *no* moustache stripes. Back and upperwing coverts are usually brown with wide white feather edges, but paler birds have white backs with brown spots. Flight feathers above are brown with heavy white spotting. White tail has brown bands but is unbanded on palest birds. Underparts are white with a variable amount of narrow brown streaking. Leg feathers and undertail coverts are usually unmarked but may show some fine short brown shaft streaks. Underwing is white with a variable amount of brown streaking on coverts and dark brown wingtips. Cere, eye-rings, and legs are blue-grey.

Unusual plumages

No unusual plumages have been reported.

Similar species

Peregrines (Plate 46) can appear similar. See under that species for distinctions.

Northern Goshawk (Plate 23) adults can appear similar to Gyrfalcons in powered flight, as their wings appear pointed but have stronger head pattern, distinctly barred breasts, and more heavily marked flight feathers.

Flight

Powered flight is with slow, deliberate, powerful wingbeats. Resulting flight is rapid and is faster than that of Peregrine. Soars and glides on level wings, but sometimes glides with wrists below body and wingtips swept up. Gyrfalcons hover briefly on occasion, particularly when looking for prey that has taken cover.

Moult

Annual moult of adults is complete; it begins in spring and is completed by late autumn. Juveniles begin a body moult in winter; they moult flight and tail feathers at the same time as adults do.

Behaviour

Prey mainly on birds, particularly Ptarmigan and Willow Grouse, but also take seabirds, ducks, and some mammals, especially lemmings. Three hunting methods are: searching from an elevated perch commanding a good view of terrain and launching out after prey when sighted; flying rapidly in a terrain hugging manner and surprising prey after coming over a hillock; and, less often, searching the ground from a high soar and stooping rapidly when sighted. Most prey is taken on ground, but can be taken in air. Gyrs possess great stamina and will tail chase until prey is exhausted. Gyrs also pirate prey from other raptors.

Carrion, including reindeer, foxes, and fish, is also eaten, most unusually for a falcon, but perhaps a necessity for one living in a harsh climate.

Gyrfalcons nest usually on cliffs or crags, using the old stick nest of other raptors or Raven, but sometimes they use old stick nests in trees. Display flight of male includes cliff racing and high soaring. Because nest areas are so remote, few people have witnessed these flights.

Gyrfalcons are usually solitary, and are vocal only during the breeding season.

Status and distribution

Rare and local sedentary breeders in remote arctic mountains or seacoasts in Iceland, Norway, Sweden, Finland, and Russia. In winter some juveniles, particularly from Russia, but also perhaps from Greenland, move south, heading for coasts and other open areas; they usually stay north of 60°.

Vagrants, usually juveniles, have been reported throughout Europe except in Mediterranean areas. Some vagrant white-morph birds (most likely from Greenland) have been reported.

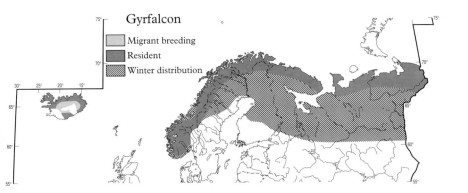

Fine points

Gyrfalcons have a white colour morph in North America, Greenland, and eastern Asia. Some birds in North America and Greenland are extremely dark; this is thought to be an extreme of the grey morph, as many birds are intermediate.

Subspecies

Monotypic. Formerly race *rusticolus* was described for Europe, but this differs not at all from Gyrs of Asia, Greenland, and North America.

Etymology

'Gyrfalcon' derived from Latin *gyrfalco* or *girofalco*, thought to be either a corruption of *hierofalco*, meaning 'sacred falcon', for its exalted place in falconry or from the Old German *gir* meaning 'greedy', for its rapacity; *rusticolus* is Latin for 'living in the country', most likely referring to its tundra habitat.

Measurements

Length: 50–65 cm (57 cm)
Wingspread: 110–130 cm (121 cm)
Weight: male 0.9–1.3 kg (1.1 kg)
female 1.3–2.1 kg (1.7 kg)

Barbary Falcon *(Falco pelegrinoides)*

Plate 46; Photos, p. 354

Description

A compact falcon, smallest of the 'large' falcons, a local inhabitant of hilly and mountainous deserts of North Africa, Sinai, and the Arabian peninsula. They are essentially a smaller, paler, desert-adapted Peregrine and, with good views, can be distinguished in the field. Adults by sex are almost alike in plumage, but females are separably larger. Juvenile plumage is different from that of adults. Adults show rufous nape, but distant falcons appear dark hooded.

Adult male. Head consists of narrow buffy forehead, dark slate crown with rufous-buffy feather edges, dark slate areas behind eyes, long narrow dark slate moustache marks below each eye that separate creamy throat from large creamy cheek patches, and *rufous nape with dark slate ocelli or 'false eyes'*. Back and upperwing coverts are slate-grey grading into blue-grey on rump and uppertail coverts. Uppertail is blue-grey with blackish bands; bands progressively wider toward tip. As a result, tail appears quite two-toned. Underparts are rufous-buff in fresh plumage but fade to creamy and are unmarked except for some short black barring on flanks and undertail coverts and sparse black spotting on lower breast and belly. Distant flying falcons appear unmarked. Underwings appear pale; *rufous-buff coverts contrast with flight feathers in fresh plumage*. Dark carpal comma and dark tips of primaries are noticeable on underwings; the latter as a narrow dark band on outer trailing edge. Wingtips reach tail tip on perched falcons. Cere, eye-rings, and legs are bright yellow.

Adult female. Similar to adult male, but rufous-buff wash and markings on underparts average heavier and colours of soft parts are duller.

Juvenile. Head consists of dark brown crown that has buffy feather edges, dark brown upper nape, areas behind eyes, and narrow moustache marks that separate buffy throat from large unmarked buffy cheeks, and rufous-buff superciliaries that extend on to mid-nape, joining there to form a 'V'. Lower nape is rufous-buff with some dark markings. Dark brown back and upperwing and uppertail coverts have rufous-buff feather edges. Dark brown tail has seven to nine narrow buffy to rufous-buff bands and a wide buffy terminal band; last two bands are narrower and wider spaced and, consequently, tail on distant falcons can appear to have darker subterminal band. Rufous-buff to creamy (faded) underparts are marked with fine dark brown streaking that is thicker on belly and flanks. Underwings are uniformly marked but nevertheless appear somewhat pale. Wingtips fall just short of tail tip on perched birds. Cere and eye-rings are bluish on fledglings but become yellow sometimes during their first autumn. Legs are greenish-yellow, later becoming yellow.

Unusual plumages
No unusual plumages have been described.

Similar species
Barbary Falcons appear much like small Peregrines in shape and behaviour.

Peregrines (Plate 46) are larger, darker, and more heavily marked, and show darker underwings and thicker moustache marks. Males are the same size as female Barbarys and have same wing to tail proportions, but female Peregrines are relatively longer tailed compared to male Barbarys. *Adults* are darker slate on head without rufous nape (except for some *brookei*, which are even darker on head and back), more whitish (darker rufous on *brookei*) and heavily marked on undersides, and have darker underwings with less noticeable dark primary tips. Two-toned coloration on uppertail less noticeable than that of Barbary adult. *Juveniles* have wider, heavier breast streaking and lack rufous-buff superciliaries that form 'V' on nape and dark subterminal tail band. Markings on leg feathers differ in being either dark chevrons, large spots, or wide streaks. Cere and eye-rings remain bluish until first summer.

Lanner Falcons (Plate 44), specifically perched *adults* of north African races, appear quite similar to adult Barbarys; both have blue-grey-backs, rufous napes, and wingtips that reach tail tip. But Lanners have rufous on crown, small blackish patch on forehead, and a more well defined narrow dark line behind each eye. Underparts are white, not creamy, and flanks are spotted, not barred as on adult Barbarys. Flight action and silhouettes are quite different. *Juveniles* are quite different, with heavier dark streaking on underparts, two-toned underwing pattern, and bluish cere and eye-rings.

Merlin (Plate 43) adult males are similar to low-flying adult male Barbarys; both have blue-grey upperparts, rufous napes, and dark terminal tail bands, but Merlins are much smaller, have heavily streaked underparts, and lack bold moustache marks.

Flight

Powered flight is Peregrine-like but, because of smaller size, with more rapid wingbeats. Soars on flat wings, often with tail fanned. Glides with wings level or with wrists below body and wingtips swept up.

Moult

Annual moult of adults is complete; it begins after breeding (June) and is completed by November. Post-juvenile moult begins in early spring (March) and is completed before the end of year.

Behaviour

Superb flyers that capture a wide variety of birds in the air, as would be expected for a falcon so similar to the Peregrine. Being smaller, they are even more agile. Prey are captured by spectacular stoops and by surprise from rapid flight. Barbarys often hunt from a high perches searching for a flying bird that is far from cover. When such is spotted, they fly rapidly directly toward it, get above it, and then stoop. Infrequently, bats are also captured in flight.

Barbarys nest on cliffs in hilly desert areas. Breeding behaviour and biology are similar to those of Peregrines, as are vocalizations.

Status and distribution

Uncommon, local breeders in hilly and mountainous deserts and semi-deserts of North Africa, the Canary Islands, Sinai, and the Arabian peninsula. Actual distribution is little known outside of Morocco and Israel.

They are most probably non-migratory, but young disperse. Some adults may leave hotter areas in summer, and during northern winter some may move south or into cities to find better food sources.

Fine points

Underwings of adults are paler than those of adult Peregrines and usually show dark carpal comma and wing tips and salmon to buffy coverts.

Subspecies

F. p. pelegrinoides is the race in the Western Palearctic. *F. p. babylonicus* of Asia is identical in plumage, with no other distinctions other than range, which is continuous from Morocco to western China. (*Note*: The taxonomic status of the Barbary Falcon is not clear; they are

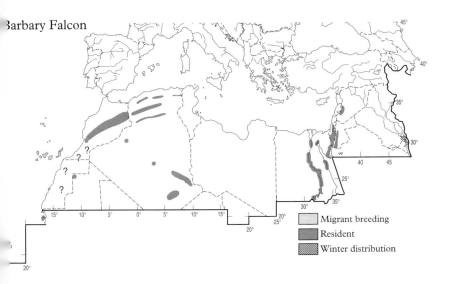

Barbary Falcon

Migrant breeding
Resident
Winter distribution

obviously closely related to Peregrine *Falco peregrinus* but differ in being desert-adapted.)

Etymology

'Barbary' after the Barbary countries of north-west Africa (Morocco, Algeria, and Tunisia), where the first specimen was taken; *pelegrinoides* means 'Peregrine-like'.

Measurements

Length: 32–40 cm (36 cm)
Wingspread: 82–99 cm (88 cm)
Weight: male 350–480 g (430 g)
　　　　　female 550–720 g (640 g)

Peregrine *Falco peregrinus*
Plate 46; Photos, p. 366

Description

A widespread, bold, robust large falcon that breeds locally through-out the Western Palearctic. Three subspecies occur: *peregrinus* is widespread in central and northern Europe; *calidus* breeds in the Arctic tundra; and *brookei* is resident around the Mediterranean, as well as Spain, Turkey, and the Caucasus. Adults of both sexes are almost alike in plumage, but females are separably larger. Juvenile plumage is different from that of adults. *All show dark 'hoods'.* Wingtips reach (adults) or almost reach (juveniles) tail tip on perched falcons. The following descriptions are for *F. p. peregrinus*.

Adult male. Head is dark slate except for white throat and cheeks separated by wide dark slate moustache mark below each eye. Back and upperwing coverts are dark blue-grey with short blackish bar-ring grading to a paler blue-grey on uppertail coverts. Flight feathers appear blackish on uppersides. Relatively short blue-grey tail has numerous black bands that are wider on lower half; tail can appear two-toned on distant birds. Underparts are white, with short black barring across belly and on undertail coverts and leg feathers, some-times with pale creamy or rufous-buff wash and short narrow dark streaks or spots on breast. Uniformly heavy marked underwings appear dark on distant flying birds. Cere, eye-rings, and legs are bright yellow to orange-yellow.

Adult female is similar to adult male except for more brownish cast to upperparts, particularly in worn plumage. Females on aver-age are more heavily marked on underparts and have darker rufous-buff wash on underparts. Cere, eye-rings, and legs are yellow, duller on average than those of adult male.

Juvenile. Head consists of pale forehead, dark brown crown and nape (often with buffy feather edging), and dark brown moustache marks that separate whitish throat from whitish cheeks, which often have some narrow dark streaks. Nape often shows buffy markings. Back and uppertail and upperwing coverts are dark brown with narrow buffy feather edges. Uppersides of primaries are dark brown. Tail is longer than that of adult and is dark brown with more or less complete narrow rufous-buffy bands and wide buffy terminal band. Underparts are creamy to buffy with heavy dark brown streaking. Heavily marked underwings appear dark on distant flying birds. Creamy leg feathers are marked with dark brown chevrons. Creamy undertail coverts have narrow dark brown markings that vary from streaks to barring. Cere and eye-rings are bluish; legs are pale yellow to yellow.

F. p. calidus is quite similar to *peregrinus*, but falcons are on average longer winged and are paler in coloration.

Adult. Head is more dark blue-grey and less dark slate, moustache marks are narrower, and underparts are whiter and usually lack the rufous-buff wash, with little or no dark breast markings; females average heavier markings. Belly is often marked with spots in centre and short bars on flanks.

Juveniles are more similar to juvenile *peregrinus* but moustache marks are narrower, heads on average paler brown, and leg feathers are sometimes marked with wide dark brown streaks.

F. p. brookei differs from other races by smaller size and darker coloration.

Adult. Head and upperparts darker, sometimes with rufous-brownish feather edges on back and upperwing coverts. Nape may have rufous patches; this is present on many birds in Spain, northwest Africa, and the Caucasus, but only occasionally on Italian and Sicilian falcons. Moustache marks average thicker; pale cheeks on average smaller. Underparts usually have a strong rufous wash; females on average have heavier markings.

Juvenile. Head has noticeable rufous-buff nape patches, and upperparts have wide rufous feather edges. Underparts have strong rufous-buff wash and wide dark streaking. Leg feathers are marked with chevrons.

Unusual plumages

Sight records exist of partial albinos. Specimens of both adult and juvenile dilute-plumage birds have pale brown upperparts and pale brown markings on underparts.

Similar species

Gyrfalcons (Plate 45) are also large dark falcons but are larger and have two-toned underwings; relatively longer, wider, more tapered tails; wider wings; and less noticeable moustache marks. Wingtips fall short of tail tips on perched Gyrs.

Lanner Falcons (Plate 44) can appear similar to same-age Peregrines. See under that species for distinctions.

Saker Falcons (Plate 45) can appear similar to juvenile Peregrines. See under that species for distinctions.

Eleonora's Falcons (Plate 41) and **Eurasian Hobbys** (Plate 42) can appear similar to juvenile Peregrines. See under those species for distinctions.

Flight

Powered flight is rapid and direct on shallow stiff, powerful wing-beats, similar to those of cormorants. Soars on flat wings with widely fanned tail. When tail is spread so that outer feathers almost touch wing, tail appears diamond-shaped. Glides with wings level or with wrists below body and wingtips swept up. Wingtips bow noticeably upwards when falcon executes high-speed turns. (For aerial hunting techniques, see 'Behaviour' in this account.)

Moult

Annual moult of adults is complete, beginning in April and completed by end of year. Arctic adults begin later, suspend moult during migration, and complete on winter grounds. Post-juvenile moult is usually complete and begins earlier than that of adult but is variable in timing and completeness.

Behaviour

Superb flyers that capture a wide variety of birds in the air, the ultimate avian predator. Their capture flights, which usually end in a spectacular stoop, are rapid and direct, with prey often struck initially but not grabbed until later. Falcons also hunt by sitting on a high perch and searching for a bird that is flying far from cover. When such is spotted they rapidly fly directly toward it, get somewhat above it, and then stoop. Sometimes prey tries to avoid capture by climbing; this results in a contest of endurance with the Peregrine also climbing. Such birds usually tire first and are grabbed by the falcons as they try to return to ground. If missed, Peregrines rarely tail chase fast flying birds such as Pigeons. Infrequently, bats are also captured in flight.

Despite many claims of faster flight, the maximum velocity of stooping Peregrines is most likely no more than 140 km/h, as determined by radar studies. Golden Eagles are faster in dives as reported by many observers.

They nest on cliffs in a wide variety of habitats, avoiding only extensive tracts of forest, and have adapted to nesting on many man-made structures, such as castles, bridges, sky-scrapers, rock quarries, and even water towers—as long as there is a flat surface resembling a cliff ledge. Some falcons have used the old tree nests of other raptors, but this is rare. Areas around nests are strongly defended against all other raptors. Fledglings often chase and capture flying insects, such as dragonflies.

As would be expected from such an aerial hunter, display flights are equally spectacular and consist of a variety of aerobatics such as cliff racing, undulating flight, high circling, and others.

Vocalizations are well known and are usually related to nesting. Peregrines bathe and drink frequently; they usually occur near water.

Status and distribution

Breed locally on coastal and mountain cliffs throughout most of the Western Palearctic. Their status varies from common in mountains of Spain, Algeria, and Morocco to fairly common in Ireland, Scotland, Wales, parts of France and Italy, southern Israel, and the former USSR to uncommon, rare, or absent elsewhere. They are absent as breeding birds from east central Europe and in areas with no cliffs, except for parts of northern Germany, where they use abandoned tree nests of other raptors.

A marked decline in most populations occurred in the 1950–70 period due to various persistent pesticides. The effects of these have diminished but many populations have not returned to former

Peregrine

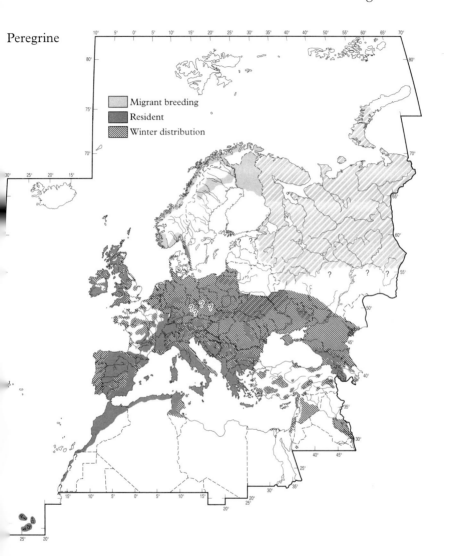

levels, particularly those in Scandinavia, Finland, and southern parts of the former USSR.

Nowadays, the main threat is thought to be from egg collectors and falconers illegally taking nestlings.

Many birds are year around residents, but some eastern and northern continental birds move to south-western Europe for winter. The entire arctic population migrates, with many going into sub-Saharan Africa. Water is no barrier to this falcon on migration; they are frequently encountered far out at sea.

Fine points

Compared to other large falcons—Saker, Lanner, and Gyrfalcon—inner wing (arm) is shorter and outer wing (hand) is longer; wrist is relatively closer to body.

Subspecies

F. p. peregrinus is resident on the British Isles and resident (west) and summer breeder (east and north) of continental Europe north to tree line and north of the Mediterranean basin. *F. p. calidus* is a summer breeder in arctic areas of Scandinavia, Finland, and Russia. *F. p. brookei* is sedentary throughout the western and northern Mediterranean basin, westward across the Iberian Peninsula, and eastward through Turkey to the Caucasus.

Etymology

peregrinus is Latin for 'wandering', for the falcon's long-distance migrations and dispersals.

Measurements

Length: 37–47 cm (42 cm)
Wingspread: 94–116 cm (104 cm)
Weight: male 580–750 g (650 g)
female 750–1150 g (950 g)

References

Articles cited in text

Frumkin, R. and Clark, W. S. (1988). Is there a dark morph of the Sooty Falcon *Falco concolor*? *Ibis* **130**, 569–71.

Harris, A., Tucker, L., and Vinicombe, K. (1990). *Bird Identification*. Macmillan, London.

Hiraldo, F., Delibes, M., and Calderon, J. (1976). [Sobre el status taxonomico del Aguila imperial iberica.] In Spanish, with English summary. *Doñana Acta Vert.* **3**, 171–82.

James, A. H. (1984). Geographic variation in the Buzzard *Buteo buteo* (Linnaeus, 1758): Mid-Atlantic and west Mediterranean islands. (Aves: Accipitridae). *Beaufortia* **34**, 101–16.

Love, J. (1983). *The Return of the Sea Eagle*. Cambridge University Press, Cambridge.

Watson, D. (1977). *The Hen Harrier*. Poyser, Berkhampstead.

A great source of information and references on the regularly occurring species

Forsman, D. (1998). *The Raptors of Europe and the Middle East*. Poyser, London.

Articles published after Forsman (1998) went to press

Clark, W. S. and Schmitt, N. J. (1998). Ageing Egyptian Vultures. *Alula* **4**, 122–7.

Clark, W. S. and Yosef, R. (1998). *Raptor in-hand Identification Guide*. International Birdwatching Centre, Eilat, Israel.

Clark, W. S. (1997). Identification of perched Montagu's and Pallid Harriers. *Birding World* **10**, 267–9.

References for vagrant species (vagrants were not covered in Forsman (1998))

Crested Honey Buzzard

Forsman, D. (1994). Field identification of Crested Honey Buzzard. *Birding World* **7**, 396–403.

Laine, L. (1996). The 'Borka Puzzle'—the first Western Palearctic Crested Honey Buzzard. *Birding World* **9**, 324–5.

Shirihai, H. (1994). The Crested Honey Buzzard in Israel—a new species. *Birding World* **7**, 404–6.

Symens, P., Gaucher, P., and Wacher, T. (1996). Crested Honey Buzzards in Saudi Arabia in October 1994. *Dutch Birding* **18**, 126–9.

Vaurie, C. and Amadon, D. (1962). Notes on the Honey Buzzards of eastern Asia. *Am. Mus. Novitates*, no. 2111. American Museum of Natural History, New York.

Pallas's Fishing Eagle

Porter, R., Christensen, S., and Schiermacker-Hansen, P. (1996). *Field Guide to the Birds of the Middle East*. Poyser, London.

van IJzendoorn, E. J. (1979). Pallas's Fish Eagle *Haliaaetus leucoryphus* of Zuidelijk flevoland. *Dutch Birding* **1**, 10–15.

van IJzendoorn, E. J. (1980). Comments on Pallas's Fish Eagle *Haliaaetus leucoryphus* of Zuidelijk flevoland. *Dutch Birding* **2**, 8–9.

Bald Eagle

Clark, W. S. and Wheeler, B. K. (1987). *A Field Guide to Hawks (of) North America*. Peterson Field Guide series, no. 35. Houghton Mifflin, Boston, MA.

Wheeler, B. K. and Clark, W. S. (1995). *A Photographic Guide to North American Raptors*. Academic Press, London.

Ruppell's Vulture

Porter, R., Christensen, S., and Schiermacker-Hansen, P. (1996). *Field Guide to the Birds of the Middle East*. Poyser, London.

Zimmerman, D. A., Turner, D. A., and Pearson, D. J. (1996). *Birds of Kenya and northern Tanzania*. Princeton University Press, Princeton, NJ.

Bateleur

Porter, R., Christensen, S., and Schiermacker-Hansen, P. (1996). *Field Guide to the Birds of the Middle East*. Poyser, London.

Shirihai, H. (1996). The Birds of Israel. Academic Press, London.

Zimmerman, D. A., Turner, D. A., and Pearson, D. J. (1996). *Birds of Kenya and northern Tanzania*. Princeton University Press, Princeton, NJ.

Dark Chanting Goshawk

Bergier, P. (1987). [Les Rapaces Diurnes du Maroc.] In French. *Annales du CEEP (ex-CROP)*, no. 3. Centre d'études sur les Eco. des Prov., France.

Porter, R., Christensen, S., and Schiermacker-Hansen, P. (1996). *Field Guide to the Birds of the Middle East*. Poyser, London.

Shirihai, H. (1996). *The Birds of Israel*. Academic, London.

Zimmerman, D. A., Turner, D. A., and Pearson, D. J. (1996). *Birds of Kenya and northern Tanzania*. Princeton University Press, Princeton, NJ.

Shikra

Clark, W. S. and Parslow, R. (1991). A specimen record of Shikra *Accipiter badius* for Saudi Arabia. *Sandgrouse* **13**, 44–6.

Labinger, Z., Gorney, E., and Parslow, R. (1991). First record of the Shikra *Accipiter badius* in Israel. *Sandgrouse* **13**, 46–9.

Porter, R., Christensen, S., and Schiermacker-Hansen, P. (1996). *Field Guide to the Birds of the Middle East*. Poyser, London.

Zimmerman, D.A., Turner, D. A., and Pearson D. J. (1996). *Birds of Kenya and northern Tanzania*. Princeton University Press, Princeton, NJ.

Verreaux's Eagle

Porter, R., Christensen, S., and Schiermacker-Hansen, P. (1996). *Field Guide to the Birds of the Middle East*. Poyser, London.

Shirihai, H. (1996). *The Birds of Israel*. Academic Press, London.

Verdoorn, G. (1988). Sexual plumage dimorphism in the Black Eagle. *Gabar* **3**, 73–5.

Zimmerman, D. A., Turner, D. A., and Pearson, D. J. (1996). *Birds of Kenya and northern Tanzania*. Princeton University Press, Princeton, NJ.

Tawny Eagle

Porter, R., Christensen, S., and Schiermacker-Hansen, P. (1996). *Field Guide to the Birds of the Middle East*. Poyser, London.

Shirihai, H. (1996). *The Birds of Israel*. Academic Press, London.

Zimmerman, D. A., Turner, D. A., and Pearson, D. J. (1996). *Birds of Kenya and northern Tanzania*. Princeton University Press, Princeton, NJ.

Amur Falcon

Corso, A. and Dennis, P. (1998). Amur Falcons in Italy—a new Western Palearctic bird. *Birding World* **11**, 259–60.

Corso, A. and Clark, W. S. (1988). Identification of Amur Falcon. *Birding World* **11**, 261–8.

Porter, R., Christensen, S., and Schiermacker-Hansen, P. (1996). *Field Guide to the Birds of the Middle East*. Poyser, London.

Zimmerman, D. A., Turner, D. A., and Pearson, D. J. (1996). *Birds of Kenya and northern Tanzania*. Princeton University Press, Princeton, NJ.

Appendix: Colour photographs of raptors

Osprey

1. **Adult.** Adults have uniformly dark upperparts. Note small white head with dark eye-stripe.

2. **Juvenile.** Juveniles show pale edges to back and upperwing coverts and orangish eyes.

3. **Adult.** Classic shape with crooked wings. Note dark carpal patches, white body, and greyish secondaries.

4. **Melanistic.** Abnormal plumage with saturation of dark coloration.

(1)

(2)

(3)

(4)

Eurasian Honey Buzzard

1. **Adult male.** Adult males have grey faces. Note pattern on upperwings and uppertail.

2. **Adult male.** Adult males have grey faces. Note clean dark wingtips.

3. **Adult male.** Adult males have 'dipped-in-ink' black tips on outer primaries and wide unbanded area on secondaries.

4. **Adult male.** Adult males have grey faces; adults have dark ceres.

5. **Adult female.** Adult females lack black 'dipped-in-ink' tips on outer primaries but do show pale primary patches on undersides of back-lighted underwings. Female secondaries are usually slightly darker than primaries and show more barring than those of adult males.

6. **Adult female.** Adult females have brown faces. Note short legs on perched birds.

7. **Juvenile.** Outer primaries of juveniles have extensive black tips; secondaries are darker than primaries. Note pale bar across underwing coverts.

(1)

(2)

(3)

(4)

(5)

(6)

(7)

Black-shouldered Kite

1. **Adult.** Adults have red eyes. Wingtips extend beyond tail tip.

2. **Juvenile.** Juveniles have orangish eyes, grey-brown on crown, and white tips on back and upperwing coverts.

3. **Juvenile.** Juveniles have orangish eyes, grey-brown on crown, and white tips on back and upperwing coverts.

4. **Juvenile.** Juveniles have orangish eyes, grey-brown crowns, and dark secondaries.

5. **Adult.** Note pointed wings and white body and tail.

6. **Adult.** Often hover with legs dangling.

(1)

(2)

(3)

(5)

(4)

(6)

Red Kite

1. **Adult.** Adults have white heads and dark breast streaks. Note deeply forked tail.

2. **Juvenile.** All Red Kites show a pale bar across each upper-wing.

3. **Adult.** Adults have white faces and dark streaks on breast. White primary panels and long rufous, deeply forked tail are diagnostic.

4. **Juvenile.** Juveniles have buffy faces and pale streaks on breast.

(1)

(2)

(3)

(4)

Black Kite

1. **Adult.** Some adults show little white on the face.

2. **Adult.** Extremely dark individual.

3. **Adult.** Typical adult.

4. **Juvenile.** Juveniles usually show narrow white line through midwing. Note the square tail tip lacking a notch.

(1)

(2)

(3)

(4)

White-tailed Eagle

1. **Adult.** Short white, wedge-shaped tail, dark undertail coverts, and parallel-edged wings are distinctive.

2. **Juvenile.** Juveniles show no moult in flight feathers. All immature White-tailed Eagles show white axillaries and bar across underwing.

3. **Second winter.** Similar to juvenile, but with much more white on body and flight feather moult obvious in secondaries as a mix of longer, paler old and shorter, darker new feathers.

4. **Juvenile.** Juveniles are overall dark, with dark beak and eyes. Note that the white on each tail feather of immatures ends in a spike. Note that tarsi are not feathered to toes.

(1)

(2)

(3)

(4)

Egyptian Vulture

1. **Adult.** Adults have distinctive black and white underwing pattern, with white body and tail.

2. **Juvenile.** Juveniles are overall brownish, with paler tail coverts. All flight feathers same age.

3. **Juvenile.** Juveniles have dark feathers on forehead forming a 'widow's peak'.

4. **Second winter.** Similar to juvenile but with tail coverts same colour as body. Note flight feather moult.

(1)

(2)

(3)

(4)

Bearded and Lappet-faced Vultures

1. **Adult Bearded Vulture.** Note distinctive shape of wings and tail. Adults have pale heads.

2. **Adult Bearded Vulture.** Note tuft of feathers under beak that form the 'beard'.

3. **Adult Lappet-faced Vulture.** Race *tracheliotus*. Face is red, and leg feathers are white.

4. **Juvenile Lappet-faced Vulture.** Juvenile *negevensis* have white heads. Note species' short tail and serrated trailing edge of wings.

(1)

(2)

(3)

(4)

Griffon Vulture

1. **Adult.** Underwings show long pale wide bar and shorter narrow pale bar on coverts. Adults have pale beak and eyes and pale secondaries with dusky streaks.

2. **Adult.** Paler upperwing coverts contrast with darker flight feathers. Adults have pale beak and short white ruff.

3. **Juvenile.** Juveniles have dark beak and eyes, long pointed rufous-brown ruff, and narrow pale streaks in upperwing coverts.

4. **Juvenile.** Juveniles have pointed tips to uniformly dark secondaries and show less obvious long wide pale bar on underwing coverts.

(1)

(2)

(3)

(4)

Cinereous Vulture

1. **Juvenile.** Overall blackish. Note serrated trailing edge of wings and wedge-shaped tail tip.

2. **Adult.** Eagle-like in flight, lacking the secondary bulge of *Gyps* vultures. Note pale head and narrow pale line at base of flight feathers.

3. **Juvenile.** Overall blackish with pale bluish face skin and cere (sometimes pinkish). Note head shape compared to that of juvenile Griffon Vulture.

(1)

(2)

(3)

Short-toed Snake Eagle

1. **Adult.** Adults have dark brown bibs and dark marks near tips of secondaries. This is most likely a male because of the white vertical streaks in the bib.

2. **Juvenile.** Juveniles have rufous-brown bibs and lack dark marks near tips of secondaries.

3. **Adult.** Pale individual with necklace of dark marks in place of dark bib.

4. **Adult.** Adults have dark brown bibs. This is most likely a female as the bib is uniformly dark.

(1)

(2)

(3)

(4)

Marsh Harrier

1. **Adult male.** Tricoloured pattern on upperparts is diagnostic.

2. **Adult female.** Adult females have pale eyes, creamy breast band, pale undersides of primaries, and rufous uppertail coverts and leg feathers.

3. **Juvenile.** Overall dark brown except for white crescents at base of outer primaries. Variant with pale nape patch.

4. **Dark-morph adult male.** Overall blackish-brown with white on undersides of flight feathers.

5. **Second-winter male.** Similar to adult female, but with greyish cast to outer secondaries and primary coverts.

(1)

(2)

(3)

(4)

(5)

Hen Harrier

1. **Adult male.** Note hooded look and black line on trailing edge of underwing.

2. **Adult female.** Adult females have creamy underparts and noticeable white bars through undersides of secondaries.

3. **Juvenile.** Juveniles have more rufous underparts and greyish undersides of secondaries.

4. **Adult female.** All plumages show large white patch at base of uppertail.

5. **Juvenile.** Juvenile females have dark brown eyes (males' eyes are grey-brown). Juveniles show narrow streaks on undertail coverts and dark undersides of secondaries.

6. **Juvenile.** Juveniles of North American *hudsonius* lack extensively streaked underparts.

(1)

(2)

(3)

(5)

(4)

(6)

Montagu's Harrier

1. **Adult female.** Wide white band in secondaries reaches body. Note lack of face ring, especially across throat.

2. **Juvenile.** Note flank streaks and dark tips of primaries. Face ring is narrow and not prominent.

3. **Adult male.** Individual with all grey underparts. Note 'pigeon-headed' appearance.

4. **Adult female.** Note short legs, wingtips that reach tail tip, and lack of facial ring, especially across throat.

5. **Second winter female.** Like adult female, except that eyes are dark and underparts have a rufous, not buffy or white, cast.

6. **Second-summer male.** Juveniles undergo a body moult during their first winter, appearing complete on this male. Note juvenile flight feathers with one new adult primary.

(1)

(2)

(3)

(5)

(4)

(6)

Pallid Harrier

1. **Adult male.** Note long legs and wingtips that fall short of tail tip.

2. **Adult female.** Tips of primaries are pale, darker secondaries contrast with paler primaries, and narrow white band in secondaries barely reaches body.

3. **Juvenile.** Note prominent face ring and lack of streaks on flanks.

4. **Second winter male.** Similar to adult male, but with noticeable hood and dusky band on trailing edge of underwings.

(1)

(2)

(3)

(4)

Levant Sparrowhawk

1. **Adult male.** Note pointed wingtips. Adult males show whitish underwings and pale rufous barred underparts.

2. **Adult female.** Note pointed wingtips. Adult females show barred underwings and boldly barred underparts.

3. **Second summer male.** Note pointed wingtips. Second summer birds are juveniles returning in the spring. They show a mix of juvenile and adult body feathers but juvenile flight and tail feathers.

4. **Second summer female.** Second summer birds are juveniles returning in the spring. They show a mix of juvenile and adult body feathers but juvenile flight and tail feathers. Eye is dark brown.

5. **Juvenile.** Juveniles before leaving the Western Palearctic in autumn show dark spots on underparts that form wide dark streaks and boldly barred underwings. Eye is brown.

6. **Juvenile.** Juveniles before leaving Europe in autumn have dark brown upperparts. Note the pale nape patch.

(1)

(2)

(3)

(4)

(5)

(6)

Eurasian Sparrowhawk

1. **Juvenile male.** Underwings and underparts are boldly barred. Note wing shape.

2. **Adult female.** Underparts are boldly barred.

3. **Adult male.** Unusual male with red eye. Perhaps an older bird.

4. **Adult female.** Nape shows two white spots.

5. **Juvenile female.** Like adult female but with rufous arrowhead-shaped markings on the breast.

6. **Adult male.** Underparts are barred rufous, cheeks are rufous, and eye is orangish.

(1)

(2)

(3)

(4)

(5)

(6)

Northern Goshawk

1. **Adult male.** Males tend to have more blue-grey upperparts.

2. **Adult female.** Females tend to have more brownish upperparts.

3. **Juvenile.** Note wide pale superciliary line and irregular dark tail bands.

4. **Juvenile.** Note wide pale superciliary line and heavily streaked underparts.

(1)

(2)

(3)

(4)

Common Buzzard

1. **Adult.** Adults have dark eyes, dark breast, barred belly, and tail with wide dark subterminal band.

2. **Juvenile.** Juveniles have pale eyes, dark breast, streaked belly, and narrow dark equal-width tail bands.

3. **Adult.** Adults have dark breast and barred belly, a pale 'U' between, and wide dark subterminal tail band.

4. **Juvenile.** Juveniles have dark breast and streaked belly, with a pale 'U' between, and many narrow dark equal-width tail bands.

5. **Whitish adult.** Overall whitish, with only dark breast and flanks and dark commas in lieu of carpal patches.

6. **Whitish juvenile.** Overall whitish, with white head, dark markings restricted to sides of breast and flanks, and dark comma on underwings.

(1)

(2)

(3)

(5)

(4)

(6)

Steppe Buzzard

1. **Adult.** Similar to adult Common Buzzard in pattern but with rufous tones on undersides and tail.

2. **Rufous-morph adult.** Underparts and underwing coverts are uniform rufous. Rufous tail shows narrow dark subterminal band.

3. **Juvenile.** Paler and more sparsely marked than juvenile Common Buzzard.

4. **Dark-morph adult.** Underparts and underwing coverts are uniform dark blackish-brown.

(1)

(2)

(3)

(4)

Long-legged Buzzard

1. **Adult.** Undersides are pale with rufous wash, heaviest on belly. Unbanded tail is white on base and rufous on tip. Note black carpals on long eagle-like wings.

2. **Rufous-morph adult.** Head, body, and underwing coverts are uniformly rufous, with black carpal patches. Belly is darker than breast. Undersides of secondaries are whitish with narrow dark banding.

3. **Long-legged Buzzards with Steppe Buzzards.** Long-legged Buzzards appear similar to Steppe Buzzards in flight but are larger and have proportionally longer wings.

4. **Juvenile.** Pale undersides, with dark carpals and belly band. Tail shows four narrow dark bands near tip. Note long eagle-like wings.

5. **Dark-morph adult.** Head, body, and underwing coverts are uniformly blackish-brown. Undersides of secondaries are dark with narrow whitish bands. Eye is dark.

6. **Dark-morph juvenile.** Head, body, and underwing coverts are uniformly blackish-brown. Undersides of secondaries are dark with narrow whitish bands. Eye is pale.

(1)

(2)

(3)

(4)

(5)

(6)

Lesser Spotted Eagle

1. **Adult.** Underwing coverts are paler than flight feathers. Note buffy tips of undertail coverts.

2. **Juvenile.** Underwing coverts are paler than flight feathers. Note heavily barred secondaries.

3. **Adult.** Upperwing coverts are paler than back.

4. **Juvenile.** Upperwing coverts are same colour as back.

5. **Second winter.** Like juvenile, but with fewer white spots. Note rounded nostril, 'stove pipe' legs, and pale tips on replacement secondaries and coverts.

(1)

(2)

(3)

(4)

(5)

Greater Spotted Eagle

1. **Juvenile.** Underwing coverts are darker than flight feathers. Note white on legs just above toes.

2. **Third winter fulvescens.** Aged by flight feather moult. Underparts and coverts dark rufous with blackish carpals and blackish streaks on belly.

3. **Adult.** Adults are overall dark brown. Note ragged nape and rounded nostril.

4. **Juvenile.** Juveniles show numerous white spots on upperwing coverts. Note rounded nostrils and 'stove pipe' legs.

(1)

(2)

(3)

(4)

Steppe Eagle

1. **Adult.** Overall dark brown except for buffy throat, blackish carpals, and greyish barring on flight feathers. Note wide dark band on trailing edge of wings.

2. **Fourth winter.** Similar to adult but with white streaks on greater underwing coverts, some retained immature secondaries, and mottled brown and white undertail coverts.

3. **Juvenile.** Uniform brown with wide pale band through underwings and on trailing edge of wings. Pale undertail coverts contrast with brown belly. Inner primaries are a bit paler than others.

4. **Juvenile.** Uniform brown with whitish tips of greater upperwing coverts and secondaries forming pale lines. Note large gape.

(1)

(2)

(3)

(4)

Spanish Imperial Eagle

1. **Adult.** Note white on upper scapulars and lesser upperwing converts.

2. **Adult.** Overall dark. Note white band on leading edge of wings.

3. **Juvenile.** Overall rufous, with no streaking on underparts.

(1)

(2)

(3)

Eastern Imperial Eagle

1. **Adult.** Overall dark blackish-brown except for straw-coloured nape and cheeks and pale undertail coverts. Underwings are somewhat two-toned: darker coverts contrast with flight feathers.

2. **Second winter.** Like juvenile but replacement secondaries are longer than retained ones, resulting in a ragged trailing edge of wings.

3. **Juvenile.** Undersides are creamy-buff with heavy dark streaks that end abruptly on belly. Dark secondaries contrast with paler inner primaries and converts.

4. **Third winter.** Appears overall mottled dark and light, but with pale undertail coverts and dark throat that contrasts with straw-coloured cheeks.

5. **Adult.** Overall dark brown with straw-coloured crown, nape, and cheeks, latter contrasting with dark throat, and white spots on middle scapulars. Note large head.

6. **Juvenile.** Overall creamy-buff, with heavy dark streaking from neck to belly. Note large head.

(1)

(2)

(3)

(4)

(5)

(6)

Golden Eagle

1. **Adult.** Overall dark brown, with golden nape, grey banding on flight and tail feathers, and wide dark band on trailing edges of wings and tail. Head projects less than half the length of long tail.

2. **Juvenile.** Overall dark brown, with variably sized white patches on undersides of flight feathers. Head projects less than half the length of long tail.

3. **Adult.** Adults and older immatures show a tawny or buffy bar across each upperwing.

4. **Juvenile.** Juveniles have tail with white base and well defined dark tip that lacks grey markings and lack tawny or buffy bar on upperwings.

5. **Perched.** Perched birds show pale leg feathers, golden nape, and long tail.

(1)

(2)

(3)

(4)

(5)

Bonelli's Eagle

1. **Adult.** Typical adult wing shape.

2. **Adult.** Above adults show a variably sized white triangle.

3. **Adult.** Paler adult showing distinct wide dark band through underwings.

4. **Second winter.** Similar to juvenile but darker rufous underparts with heavy dark streaking and dark band on tail tip.

5. **Juvenile.** In fresh plumage they appear rufous below, with narrow black band through underwings not always present.

6. **Juvenile.** In faded plumage they appear creamy below, with narrow black band through underwings not always present.

(1)

(2)

(3)

(4)

(5)

(6)

Booted Eagle

1. **Light morph.** White body and wing coverts contrast with dark flight feathers. Note pale inner primaries and dusky tip of tail.

2. **Dark morph.** Underparts are uniformly dark, except for paler undertail coverts. Note pale inner primaries and dusky tip of tail.

3. **Rufous morph.** Body and lesser wing coverts are rufous. Note wide dark band through underwings and pale inner primaries and dusky tip of tail.

4. **Faded rufous morph.** Rufous body and coverts often fade to creamy buff, but wide dark band on underwings is always noticeable.

5. **Upper sides.** When banking, they show pale bar across median coverts, white 'U' above base of tail, and white 'landing lights'.

6. **Head-on view.** When gliding, they show cupped wings and white 'landing lights'.

(1)

(2)

(3)

(5)

(4)

(6)

Lesser Kestrel

1. **Adult male.** Underwings are white and nearly unmarked. Note projecting tips of central tail feathers.

2. **Adult female.** Markings on underwing coverts are contrastingly darker than those on undersides of secondaries. Note projecting tips of central tail feathers.

3. **First summer male.** Similar to adult male Common Kestrel but markings on underwing coverts are contrastingly darker than those on undersides of secondaries. Note projecting tips of central and retained juvenile outer tail feathers.

4. **Adult male.** Grey head lacks moustache. Long wings almost reach tail tip, and central feathers project beyond tip. Note pale talons.

5. **Adult female.** Smallish head shows faint moustache and no dark line behind eye. Wings almost reach tail tip.

(1)

(2)

(3)

(4)

(5)

Common Kestrel

1. **Adult male.** Hovering. Note heavily marked underwings and distinct moustache mark.

2. **Adult female.** Underwings have rather uniform dark markings. Note dark line behind eye.

3. **Adult male.** Common Kestrels show two-toned upperwings. Rufous back and upperwing coverts of adult males show dark diamond-shaped markings.

4. **Juvenile.** Juveniles have even-width dark bands on brown backs. Wingtips fall short of wide dark band on tail. Note dark eye-line.

(1)

(2)

(3)

(4)

Red-footed Falcon

1. **Second winter male.** Like adult male but with some retained juvenile feathers.

2. **Second winter female.** Like adult female but with narrow black streaks on underparts and some retained juvenile flight and tail feathers.

3. **Adult female.** Head, underparts, and wing coverts are orange-buff. Note small black areas around eye, short black moustache, and wide dark band on trailing edge of wing.

4. **Adult female.** Upperparts are blue-grey, with blackish barring.

5. **First summer male.** Juveniles returning in spring have undergone body moult.

6. **Juvenile.** Upperparts are sandy brown. Note small black areas around eye, short black moustache, and white collar.

7. **Juvenile.** Upperwings are two-toned like those of kestrels, but note pale collar and lack of rufous coloration and wide dark subterminal tail band.

(1)

(3)

(2)

(4)

(5)

(6)

(7)

Eleonora's Falcon

1. **Light morph adult male.** Note two-toned underwing, sooty-rufous wash on undersides, white cheeks, and dark nape. Adult males have yellow cere; female cere is bluish.

2. **All dark adult male.** Overall dark grey, but with two-toned underwings. Tail banding is faint.

3. **Light morph second summer.** Has banding on undersides of retained juvenile flight feathers. Note new undertail coverts with sooty-rufous wash.

4. **Darkish adult.** Like all dark adult but with white throat.

5. **Juveniles.** Darkish light, and all dark morphs. Juveniles initially have blue ceres, eye-rings, and legs.

6. **Light morph adult.** Note long wings and tail of stooping falcon.

(1)

(2)

(3)

(5)

(4)

(6)

Sooty and Barbary Falcons

1. **Adult Sooty Falcon.** Underparts are overall sooty-grey. Adult females have yellow cere, eye-rings, and legs.

2. **Juvenile Sooty Falcon.** Orange-buff underparts are heavily streaked, with darker band across breast. Note long narrow wings and tail pattern.

3. **Adult Barbary Falcon.** Rufous-buff underparts are lightly marked. This is a female by size. Adults have yellow cere and eye-rings.

4. **Adult Barbary Falcon.** Upperparts are blue-grey. Note rufous on nape and tail pattern.

5. **Juvenile Barbary Falcon.** Buffy underparts show fine narrow streaks on breast.

(1)

(2)

(3)

(4)

(5)

Eurasian Hobby

1. **Adult.** Adults have rufous leg feathers and undertail coverts. Underparts show wide dark streaks that often extend from neck to belly.

2. **Juvenile.** Juveniles have buffy leg feathers and undertail coverts. Underparts show wide dark streaks that often extend from neck to belly. Spread tail shows narrow pale bands.

3. **Juvenile.** Juveniles have dark brown upperparts with buffy feather edges. Folded tail appears unbanded.

4. **Adult.** Head shows bold pattern of short narrow superciliary, two dark moustache marks, and creamy sides of nape.

5. **Juvenile.** Juveniles have buffy leg feathers and undertail coverts. Wingtips extend barely beyond tail tip.

(1)

(2)

(3)

(4)

(5)

Merlin

1. **Adult male.** Upperparts are blue-grey. Note pale collar and wide black tail tip.

2. **Adult female.** Upperparts are dark brown. Note narrow buffy tail bands. Wingtips fall short of tail tip.

3. **Adult female or juvenile.** Underparts are heavily streaked, and underwings are rather uniformly dark. Note faint moustache mark.

4. **Adult female or juvenile (*pallidus*).** This race is overall much paler, appearing somewhat kestrel-like.

(1)

(2)

(3)

(4)

Lanner Falcon

1. **Second summer** (*tanypterus*). First adult plumage characterized by heavier spotting on underparts. Head shows dark forehead, rufous crown, and narrow dark moustache mark and eye-line.

2. **Adult** (*tanypterus*). Whitish underparts are marked with black spots. Head shows dark forehead, rufous crown, and narrow dark moustache mark and eye-line.

3. **Juvenile.** Uppersides are dark brown. Crown is whitish on North African and Middle Eastern falcons.

4. **Juvenile.** Underwings are two-toned: dark coverts contrast with paler flight feathers. Dark streaking ends abruptly on the belly.

(1)

(2)

(3)

(4)

Saker Falcon

1. **Adult.** Upperparts are brown without pale cross-bars. Wingtips fall somewhat short of tail tip. Adults have yellow ceres and eye-rings.

2. **Juvenile.** Note pale head. Wingtips fall somewhat short of tail tip. Juveniles have bluish ceres and eye-rings.

3. **Juvenile.** Underwings are somewhat two-toned; greater coverts show dark spotting. Note white spots in long tail.

4. **Juvenile.** Underparts are heavily streaked; streaks extend over belly. Wingtips fall somewhat short of tail tip.

(1)

(2)

(3)

(4)

Rough–legged Buzzard and Gyrfalcon

1. **Adult Rough-legged Buzzard.** Adults have dark terminal band on tails. Uniformly dark flanks indicate adult female.

2. **Juvenile Rough-legged Buzzard.** Juveniles have pale eyes, wide uniformly dark belly band, and dusky tip to tail.

3. **Adult Gyrfalcon.** Underwings are somewhat two toned: coverts appear darker than flight feathers.

4. **Juvenile Gyrfalcon.** Upperparts are grey-brown. Note faint moustache marks and long tail.

(1)

(2)

(3)

(4)

Peregrine

1. **Adult.** Upperparts are dark blue-grey, paler on rump and uppertail coverts. Note wide dark moustache and darker outer tail.

2. **Adult.** Whitish breast is unmarked, belly is spotted, and flanks are barred. Dark grey head and large white cheeks suggest adult *calidus*. Adults have yellow cere and eye-rings.

3. **Juvenile.** Underparts are heavily streaked, and underwings appear uniformly dark. Note wide dark moustache marks.

4. **Juvenile.** Head appears 'hooded', with dark moustache marks, and creamy underparts are heavily streaked.

(2)

(1)

(3)

(4)

Index

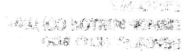